I0881630

Plains Histories
John R. Wunder, Series Editor

Also in Plains Histories

America's 100th Meridian: A Plains Journey, by Monte Hartman

American Outback: The Oklahoma Panhandle in the Twentieth Century, by Richard Lowitt

As a Farm Woman Thinks: Life and Land on the Texas High Plains, 1890–1960, by Nellie Witt Spikes; edited by Geoff Cunfer

Children of the Dust: An Okie Family Story, by Betty Grant Henshaw; edited by Sandra Scofield

The Death of Raymond Yellow Thunder: And Other True Stories from the Nebraska–Pine Ridge Border Towns, by Stew Magnuson

Food, Control, and Resistance: Rationing of Indigenous Peoples in the United States and South Australia, by Tamara Levi

Free Radical: Ernest Chambers, Black Power, and the Politics of Race, by Tekla Agbala Ali Johnson

From Syria to Seminole: Memoir of a High Plains Merchant, by Ed Aryain; edited by J'Nell Pate

"I Do Not Apologize for the Length of This Letter": The Mari Sandoz Letters on Native American Rights, 1940–1965, edited by Kimberli A. Lee

Indigenous Albuquerque, by Myla Vicenti Carpio

Law at Little Big Horn: Due Process Denied, by Charles E. Wright

Nikkei Farmer on the Nebraska Plains: A Memoir, by The Reverend Hisanori Kano; edited by Tai Kreidler

The Notorious Dr. Flippin: Abortion and Consequence in the Early Twentieth Century, by Jamie Q. Tallman

Oysters, Macaroni, and Beer: Thurber, Texas, and the Company Store, by Gene Rhea Tucker

Railwayman's Son: A Plains Family Memoir, by Hugh Hawkins

Rights in the Balance: Free Press, Fair Trial, and Nebraska Press Association v. Stuart, by Mark R. Scherer

Route 66: A Road to America's Landscape, History, and Culture, by Markku Henriksson

Ruling Pine Ridge: Oglala Lakota Politics from the IRA to Wounded Knee, by Akim D. Reinhardt

Trail Sisters: Freedwomen in Indian Territory, 1850–1890, by Linda Williams Reese

Urban Villages and Local Identities: Germans from Russia, Omaha Indians, and Vietnamese in Lincoln, Nebraska, by Kurt E. Kinbacher

Where the West Begins: Debating Texas Identity, by Glen Sample Ely

A Witness to History: George H. Mahon, West Texas Congressman, by Janet M. Neugebauer

Women on the North American Plains, edited by Renee M. Laegreid and Sandra K. Mathews

FLOOD ON THE TRACKS

Living, Dying, and the Nature of Disaster in the Elkhorn River Basin

TODD M. KERSTETTER

Plainsword by James E. Sherow

Texas Tech University Press

This book is typeset in Minion Pro. The paper used in this book meets the minimum requirements of ANSI/NISO Z39.48-1992 (R1997). ♾

Designed by Kasey McBeath
Cover photograph courtesy of Nebraska State Historical Society

Library of Congress Cataloging-in-Publication Data
[CIP TK]

18 19 20 21 22 23 24 25 26 / 9 8 7 6 5 4 3 2 1

Texas Tech University Press
Box 41037 | Lubbock, Texas 79409-1037 USA
800.832.4042 | ttup@ttu.edu | www.ttupress.org

The author gratefully acknowledges *Nebraska History* and *Great Plains Quarterly*, which published earlier versions of portions of this work.

To the Elkhorn River and those who have lived with it

CONTENTS

Illustrations xi
Plainsword xiii
Acknowledgments xvii

Introduction
Putting In: Reflections on a Plains River and Its People 3

Chapter 1
The Elkhorn River Basin 17

Chapter 2
Up the Elkhorn 35

Chapter 3
Imperfect Progress and the Elkhorn:
Flooding and Flood Fighting, 1890s–1930s 59

Chapter 4
The Flooded Forties and "The Worst Flood in History" 85

Chapter 5
Yearly Fears of Flooding: Living with the Elkhorn since 1944 117

Conclusion
Taking Out: Flooding and Climate Change on the Plains in
the Twenty-first Century 139

Notes 163
Sources 191
Index 199

ILLUSTRATIONS

Maps

I-1: Elkhorn River Basin 2
1-1: Upper Elkhorn River Basin 12
1-2: Lower Elkhorn River Basin 12
2-1: Elkhorn Valley Route, Sioux City & Pacific RR 41
5-1: Nebraska Natural Resource Districts 131

Figures

1-1: "How We Do Things at Pierce, Nebr." 6
1-2: "Night Scene on the Elkhorn near Clearwater" 8
1-3: "Ta-Ha-Zouka Park" 8
1-4: "Kids swimming in the Elkhorn" 9
2-1: Mill dam over the Logan 47
2-2: Neligh Mill, ca. 1915 48
2-3: Neligh Mill warehouse stacked with hundreds of bags of Gold Medal flour, 1911 49
2-4: C. & N. W. Depot, West Point 51
2-5: Norfolk flood, 1877 53
3-1: A view of the Elkhorn, Neligh, Nebr. 61
3-2: Neligh Mill and dam 62
3-3: Group portrait of the Neligh Mill workforce posing with bags of So-Lite flour, ca. 1911 63
3-4: Norfolk flood, 1897 64
3-5: Cutting ice 70
3-6: Norfolk flood, 1912 71
3-7: Raising dykes [sic] to hold back flood, Norfolk, Nebr., March 23, 1917 77
3-8: Man fishing west of dam, Neligh, Neb. 78
4-1: Norfolk flood, May 12, 1944 92
4-2: Norfolk flood, 1944 93

Tables

4-1: Dollar Cost of Flooding, Norfolk, 1944 93
4-2: North Fork Peak Flow during Notable Twentieth-Century Norfolk Floods 95
4-3: Costs and Benefits of Comprehensive Plans for Flood Control on the Elkhorn River 114

PLAINSWORD: THE UPS AND DOWNS OF LIVING WITH THE ELKHORN RIVER

Why do humans continually insist on building homes, farms, factories, and cities in a flood plain? It would seem that human experience has demonstrated the folly in doing this. Todd Kerstetter's *Flood on the Tracks*, an environmental history of the Elkhorn River in Nebraska, provides insights as to why humans might think building in a flood plain a good thing to do in the first place.

A good environmental history will make a reader ask if the story he/she is reading is unique to a particular place and time, or has a greater relevance, provides a window to understanding the larger world of humans living in their surroundings. On first blush, Todd Kerstetter's history of the Elkhorn River Basin might seem confined to one locale. His is a story of a river confined entirely within the High Plains. The river rises in the western portion of Nebraska, in the sand hills, and flows eastward to where it joins the Platte River just west of present-day Omaha. Over the years that Americans have occupied the Elkhorn basin, the river flooding has created a host of problems for its human occupants. But Kerstetter's narrative, while providing a vivid story of flood destruction and mayhem in the valley, imparts a story laden with a much deeper meaning than damage to human endeavors.

For six years, I served as a city commissioner and mayor of Manhattan, Kansas. I had to deal with the many of the same issues raised in Kerstetter's work, especially why would anyone build in a flood prone area? The city founders had built where once the Blue River joined the Kaw, and over the course of more than 150 years, at least four major floods swept through the commercial district of the city until the Army Corps of Engineers built a levee system around the downtown. In the western portion of the city, Wildcat Creek winds its way from north of the city to join the Kaw River, which forms the southern boundary of the town. At least twice while I was in office, Wildcat Creek flowed out of its banks and flooded apartments, a trailer court, and businesses. The inundated property owners blamed up-stream property development for their watery travails.

While researching the history of Manhattan, I asked for a collection of photographs from the engineering department. One aerial photograph of the western portion of the city caught my attention. It showed Wildcat Creek in flood stage in the early 1960s, well before any up-stream development had occurred, or any apartments or shopping centers had been built where the creek had inundated scores of acres around it. In short, developers later built a dense assembly of apartments, businesses, and a trailer court along a creek where flooding was simply a part of its nature. That photograph redefined a committee's assumptions about flooding along Wildcat Creek. Wildcat Creek did not cause the property damage suffered by the people living near its banks. The builders and developers of the property inflicted the damage upon themselves by building where they had ignored the nature of the stream.

When I reflected on Kerstetter's story of how people built their communities, factories, and farms along the Elkhorn River, and somehow expected the river to behave itself and never overflow its banks, he made sense of the flooding along Wildcat Creek. Simply, people bore the blame for these so-called "natural disasters." Moreover, anywhere river flooding causes property damage, Kerstetter's work will call into question the root cause of the losses. Clearly Kerstetter's Elkhorn River story is not an isolated tale. He provides insights as to why people would build in a flood plain, and what consequences arose as a result. His work is a lens through which the history of any river, or even Wildcat Creek, might be viewed more clearly.

Paradoxically, the effort people put into controlling the Elkhorn River only seemed to make matters worse. Kerstetter illustrates this point well in showing the unintended consequences resulting from the levee systems that Army Corps of Engineers built in an effort to protect the cities along the river. The levees did this, but they also worked to constrict high river flows, which generated greater velocity to the downstream current. This resulted in streambank erosion, and greater flooding and property damage below the levee system.

Kerstetter is quick to point out that while levees protected the high valued urban properties, the enhanced flooding caused by the levees swallowed the lower-valued properties downstream. I saw the same results play themselves out in Manhattan, Kansas, where the levee protecting the commercial downtown and residences failed to shield those downstream. I also suspect that a similar history plays itself out wherever the Army Corps of Engineers has built its flood control levees.

Are Americans afflicted with some sort of hubris when it comes to living by rivers? Kerstetter presents compelling evidence that leads to an indictment. There is an attitude among the people living in the Elkhorn River Valley that somehow they can manage the river, and in so doing improve their lives. But the Elkhorn River has a nature of its own, one that confronts and resists human efforts to subdue it. Kerstetter makes this clear by saying that "the settler colonists who came to the Elkhorn Basin in the nineteenth century planted a culture dependent upon a river it intended to tame, but could not, bequeathing to future generations a legacy of flood damage that cannot be attributed to 'natural' disaster."

The first Americans to arrive in the valley who built homes and communities viewed the river as a commodity and source of power. It's not clear that they recognized the river as a living entity, one with its own nature supporting a rich, and diverse community of plants and animals. The early settlers saw "the music of the lathe and spindle" and "gemmed" cities, ones bright and busy with commerce. But beginning in 1873, floods would show the residents that there was more to living in the valley than erecting homes and businesses. And no matter how hard they tried, the residents, even provided with resources from the state and federal governments, could never fully effect complete control over the river.

Kerstetter reveals how "people chose a close relationship to the Elkhorn and its floodplain, and those flood damages deserve to be called man-made." Until recently, people in the valley regarded floods as a "disaster", a menace to the well-being of the state. The river, so the thinking went, had to be "engineered" so people could live and prosper beside it. But the Elkhorn River, as Kerstetter so vividly shows, continued to seek its own destiny regardless the dams, levees, and flood channels designed to control it. It still flooded and destroyed what humans thought could be built safely beside it. In a testament to the ineffectual effort of humans to control it, between 1936 and 1962, economic losses due to flooding actually increased each year.

Slowly, it dawned on the people in the valley that perhaps living "with" the river might serve them better than "controlling" and "dominating" it. Especially after the devastating flood of 1993 that resulted in huge destruction throughout the entire Missouri River watershed, the federal government started encouraging a new approach toward living "with" rivers. In the Elkhorn Valley, slowly homes and property in the flood prone areas of the basin were purchased, and the land left free to flood. This resulted in a healthier river as riparian woodlands re-emerged that served to filter the water and helped to slow the velocity of high river flows. An additional benefit arose as wildlife began returning to the valley.

Kerstetter's story is one of humans beginning to take a new approach toward living in the valley. People were coming to the conclusion that a better relationship with the river would arise by not building in the flood plain in the first place. There emerged a recognition that the river had a nature of its own despite whatever anyone else might think, and that perhaps floods did serve a beneficial role for humans. Scientists began recording restored soil moisture, refilled reservoirs, and recharged aquifers depleted by center-pivot irrigation as a result of unrestrained flooding.

Kerstetter adds one other important insight to the meaning of living with the Elkhorn River. Try as hard as it may, the river cannot escape the reality that as Donald Worster points out, "the whole world in now humanized." Kerstetter recognizes how the Anthropocene may have been playing itself out in the Elkhorn River Valley. The Anthropocene may be defined as the point in time when human interactions with their environments have resulted in planet-scale effects. Clearly, Kerstetter shows humans playing a significant role in modifying one of the planet's life-supporting systems. In this regard, as some ecologists now contend, humans are a keystone species in their environments.

While Kerstetter thoroughly understands the historical "dialogue" between humans and the Elkhorn River, he also incisively notes that the river really has no mind of its own. A river does what a river does. It's humans who are "aware," "conscious," of their actions in the valley. The Elkhorn, Kertstetter astutely asserts, doesn't care about washing out roads or railroad tracks, or mill dams. More than an environmental history of river and its human inhabitants, Kerstetter's work is an important, insightful morality tale: It's time for people to remove the trappings of civilization out of flood plains and thereby tame at least one unnatural disaster.

James Sherow
Kansas State University,
Manhattan, Kansas

ACKNOWLEDGMENTS

Many people have contributed generously to this project and deserve thanks. John Wunder heads the list. John helped me find this project—twice—and guided its development both times. He's been a great mentor and friend and I'm grateful to have found my way to him. Joanna Conrad at Texas Tech University Press stuck with this project and encouraged it over several years. She sent the manuscript to two perceptive and skilled referees, who provided insightful, informed, and precise constructive criticism that helped me immensely in improving the book. I don't know who they are, but I'm grateful for their help and I thank them. Copyeditor Barbara A. Kohl did fine work polishing the manuscript. Mike Amundson has been a role model and inspiration to me since 1990. He's always been a few steps—or more—ahead of me professionally and I've gained a great deal by watching him in action.

On the research trail, I benefited from timely and friendly help from a variety of guides and Sherpas. The staffs at the University of Nebraska at Lincoln libraries and at the Nebraska State Historical Society (NSHS) helped me find things many, many times. Special thanks to Mary-Jo Miller at the NSHS for her help procuring photographs and other images. In Norfolk, Nebraska, Jeanna Ganskopf and the archives volunteers at the Elkhorn Valley Museum shared their cache of images and newspaper clippings. Dennis Smith, Norfolk's city engineer, gave me a crash course on the city's relationship to the Elkhorn and the North Fork. At TCU's Mary Couts Burnett Library, Robyn Reid helped me locate valuable items. The crack staff at Interlibrary Loan put most of those things in my hands so quickly I sometimes couldn't keep up. June Koelker, dean of libraries, generously assigned me a research office made possible by a gift from the J. E. and L. E. Mabee Foundation.

TCU provided remarkable institutional support beyond the library. Andrew Schoolmaster, dean of the AddRan College of Liberal Arts, provided two Mid-Career Summer Research grants. Peter Worthing and Jodi Campbell, as chairs of the history department, provided encouragement and course releases to compensate for the demands of supervising graduate students. Speaking of graduate students, I've had the great fortune to work with some of the brightest young scholars in the business. Jamalin Harp and Brett Dowdle helped me locate newspaper accounts of Great Plains flooding since 1945. Brett and other members of our "Writer's Roundup," Lisa Barnett, Kendra DeHart, Reilly Hatch, and Joe Schiller, read and critiqued an early version of the manuscript. Their keen minds and sharp eyes helped me to improve the prose and avoid embarrassing mistakes. Charlotte Hogg of TCU's English Department organized summer writing boot camps and affiliated year-round programs that provided inspiration, support, and accountability for me and other writers in AddRan College.

None of this could have come to pass without support on the home front. My parents, Joyce and Ned Kerstetter, infected me with the education bug and have nurtured it—and me—for my entire life. My dad also provided a helpful critique of the manuscript. Holly McFarland, my life partner, and our daughters Nora and Leah encouraged me, listened to a lot of stories about flooding and rivers, provided delightful diversions, and are ready for me to move on to another project. I am grateful to have them in my life and I appreciate their support.

Thanks to everyone. I hope the result in some way rewards the time and energy you contributed.

FLOOD ON THE TRACKS

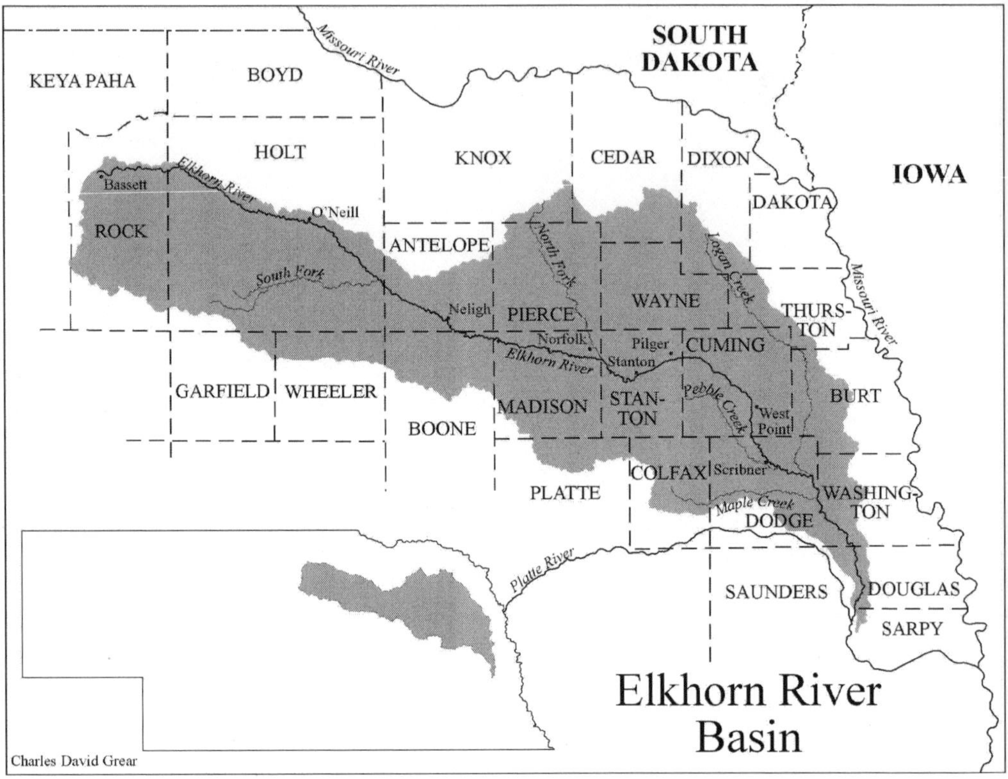

Map I-1: The Elkhorn River Basin. Map by Charles David Grear.

Introduction

PUTTING IN: REFLECTIONS ON A PLAINS RIVER AND ITS PEOPLE

This book tells stories about living with the Elkhorn River, a waterway wholly within the Great Plains, which makes it by definition a Plains river. The book might be considered an eco-biography of the Elkhorn, although the sources and nature of history dictate that it be told largely from the perspective of the people who have inhabited its banks. Most of the action unfolds at the bottom of the Elkhorn River Basin (Map I-1), a region defined by topography and gravity—the basin is the area in which water finds its way to the lowest point where it aggregates in the river. Humans also seem to have flowed downhill in the basin as much of their activity in the region has taken place in close proximity to the river. Therein lies a problem: the Elkhorn is an unreliable neighbor and cannot be counted on to stay in its place. When the volatile Great Plains environment, known best for its aridity, produces moisture, which it sometimes does in dramatic fashion, it finds its way down to rivers, which climb up from their low-lying beds and take up residence on their floodplains where, increasingly over the past two centuries, humans have also taken up residence and have been none too pleased with the river's encroachment into what they see as *their* place. This book asks the question, "What can the relationship between a Plains river and people who live on the Plains tell us?"

The book's title was inspired by Bob Dylan's album *Blood on the Tracks*. If Dylan's album deals with "the ups and downs of mature romantic relationships," as one critic suggested, this book deals with the ups and downs of living with a river in an evolving relationship rooted in romance, punctuated with tragedy, and marked by long periods of indifference. Whether that relationship could be considered mature is up for debate, but it continues to evolve. Practically speaking, the Elkhorn has for centuries touched all manner of human "tracks," including cultivated fields, camps, lodges, homes, villages, cities, paths, roads, and railroad tracks. Metaphorically speaking, songs from *Blood on the Tracks* like "Idiot Wind" and "Shelter from the Storm" offer tempting possibilities for framing the stories found here. But of all the magnificent poetry in the album's lyrics, the most appropriate for this project come from the last song, "Buckets of Rain," and go like this: "Buckets of rain/Buckets of tears/Got all them buckets comin' out of my ears." Water that feeds the Elkhorn sometimes arrives by the bucketful and when it does buckets of tears often follow.[1]

A lot of flood stories unfold in the pages ahead. In fact, it might seem like one damned flood after another. That's by design and I think anyone who has lived in the Elkhorn Basin for even thirty or forty years, a long time by human standards but a blink of an eye in the life of a river, would recognize a certain pattern or rhythm related to the river's rise or, at the very least, an awareness of the river and the possibility that it might rise. For all their similarities, the floods have differences and they expose different things at different times. By the book's end, it will be apparent that both the river and the people who live along its banks have arrived at different places, literally and metaphorically, than they occupied at the beginning and the people, at least some of them, live in a more considered, considerate, maybe even wiser way with the river.

The human regime currently ensconced in the Elkhorn's basin has been there for more than a century. It no longer clings to or depends upon the Elkhorn as it did in the late 1800s; but, the river remains a fact of life and sometimes a very unpleasant one. The unpleasantness stems from one of the traits coded in the river's DNA: flooding. The people who have lived with the Elkhorn since the mid-1800s haven't exactly been good listeners in their relationship to the river and have often been entirely too stubborn in trying to impose their will upon the river. The will of the people to dominate the river and its landscape might be said to be coded in their cultural DNA, which tells them nature can or even should be bent to serve human interests and that "imperfect" landscapes can be improved or perfected. Elkhorn floods that touched cropland, businesses, homes, roads, and, yes, railroad tracks punctuate this story of a river and its people.

Encountering the Elkhorn

I first encountered the Elkhorn during a cross-country bicycle tour in the summer of 1989. The Elkhorn Valley made me very happy for reasons I didn't fully understand or appreciate at the time. On a glorious June day, my riding buddy and I made 112 miles from Omaha to Norfolk through the lower Elkhorn Valley. We covered more miles than on any single day since setting out from Myrtle Beach, South Carolina. We had planned to camp about fifty miles out of Omaha, but the road rolled by so easily we just kept pedaling and, to our delight, ended up logging a "century," a ride of 100 miles. The next day, we made seventy-six miles from Norfolk to O'Neill. That easily beat our previous two-day record for mileage. It turns out the Elkhorn Valley had been a great highway for generations. The Pawnee and Omaha Indians traveled the valley as did settlers from the United States and abroad in the era of overland wagon trains. Railroads came up the valley not long after that and paved roads for automobiles and trucks followed soon after. In 1989, US 275 made bicycling up the valley a piece of cake. Since the last railroad that ran the length of the basin abandoned the valley in 1992, the highway has become an increasingly critical artery for transportation and commerce.

When we pedaled out of the valley after three days, a feat made possible by paved roads and other astonishing transformations of the valley during the previous century or so, I had not the slightest inkling that I would return a couple years later as a budding historian. By then, I had become interested in environmental history and John Wunder, my mentor at the University of Nebraska and editor of the Plains Histories series, suggested I consider the unexpected and largely unstudied concept of too much water in a region widely considered under-watered. That study resulted in a master's thesis that chronicles flooding in the Elkhorn Basin through 1944. Although competent for a thesis, the work left much to be desired. It notes major floods, discusses flood control, and provides limited and crude interpretation.[2]

When I finished that project I had not the slightest inkling that I would return to it a couple decades later as a somewhat seasoned mid-career historian. Re-opening what had become for me a cold case led to the discovery of new materials and this version contains new research throughout. Returning to the project allowed me to flesh out stories and to include new ones. All the material on post-1944 floods is new and the massive flood of 2010 demanded not only inclusion but also reconsideration of earlier conclusions. Reading more than two decades' worth of environmental history written since I dropped the project led me to see the Elkhorn and its sons and daughters differently and to write a more nuanced history. Historians often say every generation writes its own history. I've been thinking about the Elkhorn and studying history long enough now that this might count as a second-generation interpretation of the river, its people, and the floods they've shared. I doubt that I'll be willing or able to write a third-generation study when another couple of decades have passed, but perhaps this work will be a steppingstone for someone else to undertake that project in the future. Neither the floods nor the people seem to be going anywhere.

Other people have been encountering the Elkhorn for centuries, many on a daily basis over a significant portion of their lives and the river has left its mark on them, too. The Elkhorn, and perhaps it can stand in for other Plains rivers on this count, served as a romantic psychic center of the basin for people who lived there. Four images, three picture postcards from the first half of the twentieth century and an undated photograph, support this claim. One image produced in about 1910 shows two men in a boat, presumably on the Elkhorn's North Fork, trying to land an enormous fish (see Figure 1-1). Comparing the fish to the grown men in the boat, the fish must be about fifteen feet long. The man who hooked the fish hangs on for dear life and his companion has raised a rifle or shotgun to finish the unfortunate giant. In bright red ink the card boasts, "How we do things at Pierce, Nebr." The scene has been faked,

Figure 1-1: This photo postcard from 1910 suggests an abundance of water and fish in the Elkhorn River Basin. Author's collection.

of course, by a photographer using the day's equivalent of Photoshop. In terms of genre, the item certainly is a closer relative to postcards depicting the mythical jackalope than it is to documentary photography. But, the card conveys more than a humorous boast. It tells the viewer that Pierce is well watered, thanks to the North Fork of the Elkhorn, suggests abundance, in the form of the giant fish, and celebrates with swagger the competence of humans to subdue a giant fish under a giant sky that threatens to swallow them all. Nope, these robust Plainsmen will take care of that fish and do just fine out there on the Plains, thank you very much.

Another postcard from about the same time, "Night Scene on Elkhorn River, Clearwater, Nebraska," uses large trees to frame two people floating in a canoe on a serene river lit by a full moon (see Figure 1-2). The Elkhorn looks more like a lake than a river. The image suggests a romantic moment, to be sure, but also a welcoming landscape watered by a beautiful river with abundant timber along its banks. The picture flies in the face of the Great Plains' reputation for treelessness and probably does so intentionally. In all likelihood the image was intended not for local consumption but for export, to be mailed to someone elsewhere, probably far away from the Elkhorn Basin. For that remote consumer, the image depicts tranquility and a wealth of critical resources—water and timber—that on the Great Plains reside in river basins. It takes little imagination to project a bright future for the generation that might issue from the tryst depicted in the scene. Likewise, it takes little imagination to see that future depicted in an image from three decades later depicting Norfolk's Ta-Ha-Zouka Park in about 1941 (see Figure 1-3). Located a few miles downstream from Clearwater, the park appears in full color with a lovely blue Elkhorn River in the foreground and trees and a tranquil home located on its banks.[3]

A fourth image, an undated black-and-white photograph showing children playing in the Elkhorn as several adults supervise from the bank, comes from the collection of Emanuel Wolfe, a merchant and professional photographer who lived and worked in the Elkhorn River community of Neligh, Nebraska, in the late 1800s and early 1900s. The photo preserves an idyllic summer moment for fifteen or so children enjoying themselves in a river so wide and calm it could be mistaken for a shallow pond. A toddler, ankle deep, has stopped to assess the scene. A few feet away a girl and a boy, probably not older than ten, stand in waist-deep water. On the bank, three women, presumably the children's mothers, supervise closely but calmly. Large trees on the far bank

Figure 1-2: This picture postcard from about 1911 represents water, timber, and perhaps moon-lit romance in the Elkhorn Basin. Courtesy of Nebraska State Historical Society.

Figure 1-3: This picture postcard from 1941 celebrates tranquility, domesticity, and recreational opportunities on the Elkhorn's banks. Author's collection.

Figure 1-4: The Elkhorn could be a safe, cooling place for women and children to gather and play, as captured in this Emanuel Wolfe photograph, probably made between 1905 and 1925. Courtesy of Nebraska State Historical Society.

provide a backdrop and obscure the vast Plains sky that would dominate the view if the photographer changed his perspective ever so slightly. The Elkhorn presents no danger here. To the contrary, the docile river provides a cooling, fun, safe play space suitable for women and children.[4]

These images preserve for us what Elkhorn Basin residents in the early decades of the twentieth century wished to promote to outsiders, including a fond record of an enjoyable summer day. They illustrate the idea that Great Plains river bottoms represent "eastern" landscapes extending well into the West. The river figures prominently in each image, as do themes of abundance, tranquility, recreation, and even romance. The Elkhorn is the center, the sustainer, even the giver of life.

The literary record also points to the Elkhorn as a powerful force in people's lives. In *Sons of the River: A Nebraska Memoir* of growing up in Ewing, Nebraska, in the upper Elkhorn Basin, during the middle part of the twentieth century Norm Bomer uses the Elkhorn as an organizing device even though the river itself appears infrequently in his narrative. "A river flows through my life," Bomer writes. "Year after year, the Elkhorn River cuts a channel deep within my soul, shifting and shaping me like its own sand." This despite, as Bomer points out, the fact that the Elkhorn ranks not among the mightiest, widest, or longest rivers and that it has never irrigated his land or watered his crops.[5] Still, for Bomer and presumably the others to whom he obliquely refers in his memoir's title, *Sons of the River*, the Elkhorn not only occupied center stage, but also played the role of father figure shaping generations of boys who grew up near its banks.

That watery father figure, though, has its quirks. Living near the Elkhorn and researching its history, Bomer found that floods made strange things happen. They could, and did, push the river into new channels in a remaking of the landscape, a characteristic common to many Great Plains rivers. Floodwaters working new territory could unearth startling things such as human remains. One winter, children ice skating—one of the many recreations the river provided—found the skull and bones of an early white pioneer that had been washed out of their resting place by a flood the previous spring. Spring floods also kept students who lived across the river from the local schoolhouse from attending. Two entrepreneurial souls near Ewing operated a ferry on such occasions. For them, the Elkhorn's floods didn't disrupt commerce as much as they created it.[6]

The Elkhorn's most famous son, Johnny Carson, grew up in Norfolk, downstream from Norm Bomer's Ewing, and seems to have carried the river with him to television superstardom as the long-time host of the *Tonight Show*. As children, Johnny and his brother Dick explored and played along and even above the Elkhorn River. The Carson brothers would swing from the timbers of a railroad trestle they called "Black Bridge" that spanned the Elkhorn while trains passed over their heads. When Carson returned to Norfolk to film an autobiographical television special "Johnny Goes Home" in the fall of 1981, he decided to re-enact the stunt for his film crew. Carson spent $5,000 to hire a twenty-six-car Union Pacific train to pass over the bridge while he dangled from the timbers. As the twentieth car passed over the bridge, Carson lost his grip and dropped—all of six feet—into the chilly, shallow water of the Elkhorn. When Carson exclaimed, "It's cold!", one of the crew members recognized an opportunity to set up Carson for one of his signature bits by responding, "How cold is it?" The group broke up in laughter.[7] The river meant both playground and peril for kids growing up nearby. Even those who left, like Bomer and Carson, carried a little bit of it with them in their memories of home. The Elkhorn is a central organizing force for life in the basin and it makes strong impressions on the people who encounter it even for a short time. For basin residents, the Elkhorn is a vital part of their home on the Plains.

Region, Water, and Historians

Historians have written extensively and impressively about the world's great rivers, especially as environmental history has gained momentum in recent decades.[8] Historians of rivers in the United States have tended to write about major rivers and to focus on portions of those rivers that flow through or near major metropolitan areas. Perhaps no river-city combination has received more attention than the Mississippi River and its highly engineered relationship to New Orleans, although the relationship between Los Angeles and the Los Angeles River has been brilliantly explored, too.[9] Issues of flooding, flood control, and the alteration of the riparian environment in relation to flood control dominate most of these works, especially the ones about New Orleans and Los Angeles.

But the American West and its sub-region the Great Plains have typically received different treatment when it comes to water issues, largely because aridity overshadows most everything else. Water storage, irrigation, and drought tend to dominate the historiography.[10] For the most part, those themes extend to treatment of rivers, too, including the Missouri, arguably the most important Plains river, the Pecos, and the Brazos.[11] A long-term study of a Plains river such as the Elkhorn, which begins and ends within the region, and focuses on problems created by too much water offers an opportunity to fill gaps in the historiography.

Mention of the Great Plains calls many things to mind: seas of grass, vast skies, open flat spaces. Some might envision fields of wheat or corn, maybe ranches, bison, desolation, and wind. Water usually doesn't spring to mind, except maybe in its scarcity. Historian Walter Prescott Webb famously defined the Great Plains as a region marked by level land, a *lack* of trees, and rainfall *insufficient* to sustain the type of agriculture practiced where a humid climate prevailed and where nineteenth-century non-Indian settlers of the Plains originated. "The climate is sub-humid," Webb wrote. The entire region suffers from a "deficiency in the average amount of rainfall." According to Webb, "In this deficiency is found the key to what may be called the Plains civilization."[12] The Great Plains does receive rainfall (and snowfall and other precipitation, it should be added), sometimes in dramatic amount and fashion. Whatever the region's annual or normal precipitation, the extremes can be spectacular.[13] When it does rain, the rain can come in buckets and when it hits the ground it heads downhill to the rivers.

Great Plains rivers are an unloved and underappreciated branch of the family of rivers. Consider the US National Wild and Scenic Rivers System. Created in 1968 to protect rivers with "outstandingly remarkable" values, whatever that means, and to preserve the free-flowing condition of selected rivers, barely extends to the Great Plains. To date, only portions of three Plains rivers—the Upper Missouri in Montana, the Missouri between South Dakota and Nebraska, and the Niobrara in Nebraska—have received wild and scenic designation. Canada created a similar program, the Canadian Heritage Rivers System, in 1984. As of this writing, only a portion of the Red River in the prairie provinces, where the Great Plains extends into Canada, has received Heritage River designation. Or consider that in the massive *Encyclopedia of the Great Plains* the section on "Water" contains no entries dedicated to a river, not even the Missouri, and the "Rivers" entry occupies less than a page. Scholarly neglect notwithstanding, Great Plains rivers whether major, such as the Missouri or the Platte, or minor, such as the Republican or the Elkhorn, are central to the region's history.[14]

The Great Plains, "treeless" and "insufficiently" rained upon, made surface water—streams and rivers, as opposed to groundwater such as found in aquifers—particularly important to humans, especially before the twentieth century. Where most Plains rivers could not support navigation, river valleys provided travelers and early settlers natural overland highways. Oases of accessible water, whatever timber existed, and forage, Great Plains river valleys provided critical resources as well as shelter from the wind and cold. On the Plains, river bottoms are exceptional landscapes that could be considered "essentially veins of eastern landscape intruding far into the west."[15] When railroads and highways came into the region, they, too, often followed those natural highways. In a remote area with little wood for fuel, rivers provided energy to power mills that processed grain and other products. They have provided and continue to provide recreation ranging from fishing to tubing to boating. Rivers have always been important on the Plains.

The Elkhorn River originates and ends within the boundaries of the Great Plains, making it a "pure Plains river" and thus a superb case study for the Plains Histories series.[16] The Elkhorn, conveniently for the sake of analysis, also begins and ends entirely within the limits of the state of Nebraska. It flows into the Platte, which in turn empties into the Missouri, which flows into the Mississippi. Considered in those terms, the Elkhorn is at the end of the line, a relatively small river, a minor river, that can't support the big work of navigation that has led people to identify its relatives the Missouri and the Mississippi as important rivers, major rivers, that have captured much more attention.

For a relatively small river, the Elkhorn has a rich and lively record of interaction with humans. People lived in its basin well before the United States or Nebraska came into existence, although we don't know a lot about them. More recently in the river's long history, it lured non-Indians to settle near it, probably too near it. The natural highway along its banks facilitated overland travel up the valley and the river's current promised energy to power mills. The anomalous availability of shelter, forage, and timber beckoned. In the second half of the nineteenth century the vanguard of the United States' expansionist settler colonialism made its way up the valley and planted settlements as it went. Over time, the river also provided recreation ranging from swimming to ice skating, and destroyed things. In any number of ways the Elkhorn made life possible and at once interesting and bothersome just because it did what rivers do, which is to flow and, occasionally, overflow.

Charting a long-term microhistory of a Plains river yields insights not only to the region's environmental history, but also to the nature of disaster. A brief detour to the Mississippi River illustrates this concept. How should we interpret an event such as a flood? As an act of nature? An act of God? Do humans play a part, or are they simply victims of larger forces beyond their control? Environmental historian Ted Steinberg uses the Mississippi River flood of 1993 at Hannibal, Missouri, as a case study to consider such questions. Hannibal, Mark Twain's hometown, sits on the Mississippi's bank and the first major recorded flood there happened in 1851. By 1962, the city had had enough and asked the US Army Corps of Engineers (USACE) to design a flood protection system. The Corps proposed a project that would have protected not only historic downtown Hannibal (including Twain's house) but also outlying, and poorer, neighborhoods such as the South Side and Bear Creek Bottom. The city balked at the project's cost and appearance, though, and decided not to proceed. After a massive flood in 1973 inundated downtown, though, city leaders again asked the Corps for a flood protection plan, but a cheaper one, one that would protect downtown and Twain's home, but not the poorer sections of the city. The city moved forward with the resulting plan, a 3,000-foot long, 12-foot high, $8-million flood wall, completed in the spring of 1993.

Months later the flood hit. Downtown stayed dry, but the South Side found itself under nearly eight feet of water.[17]

Levees offer a degree of protection, but they're not perfect. Some people argue that they actually contribute to the destructiveness of high water. If levees are built on both sides of a stream, they force water to rise, rather than spread out, during heavy flow. In the event of a breach or an overtopping of the levee, water can hit the "protected" area with greater force and destruction than would have been the case without the levees. If a levee is built on one side of a stream, it might protect the area behind the levee but force water higher and with greater force onto land on the unprotected bank, thereby causing greater damage than would have occurred without the levee.

Building levees, and this principle applies to a number of other environmental modifications aimed at flood control, forces people to make choices. "We realize that the floodwall does not protect everyone, and that is indeed regrettable," the *Hannibal Courier-Post* wrote after the flood. "But wherever a line is drawn, someone always will be just on the wrong side of it." Shellia Todd, one of those wrong-side people in Hannibal, told a reporter, "We can't afford to live anywhere else but down here where it floods. It's always the poor people that get screwed."[18]

As Steinberg sees it, calling a flood such as the one that hit Hannibal in 1993 a natural disaster or an act of God or even a freak accident misrepresents reality. Natural calamities often "do not just happen," he writes, "they are produced through a chain of human choices and natural occurrences." It's a mistake to view, as we have come to do, natural disasters as "random, morally inert phenomena … beyond the control of human beings." It's also a mistake to villainize nature. People made choices about how to live in the world and while they might not have been "architects of destruction," they were certainly accomplices.[19] A long-term case study of the Elkhorn Basin reveals how people chose to live in the world of a Great Plains river.

Legal scholars Christine A. Klein and Sandra B. Zellmer apply Steinberg's principles on a grand scale in *Mississippi River Tragedies: A Century of Unnatural Disaster*. They find not only unnatural disasters, but also an unnatural river. "In truth, it's a stretch today to describe the river as anything close to 'natural,'" they write. For more than a century people have "straightened, channelized, dammed, and rip-rapped the Mississippi" to keep it *in* its channel and *out* of its natural floodplain. Humans have also created levees, floodwalls, flood insurance, and disaster relief, which have had the combined effect of luring people and property into the floodplain. As more people and property have come into the floodplain, they have inspired increasingly drastic measures to protect them from flooding. These modifications have cut off or degraded the river's ability to access its floodplain, which can be thought of as a natural relief valve for excess water. These human actions have prevented or mitigated some flood damage, but in other cases they have *exacerbated* damage stemming from natural hazards.[20]

Many of these principles played out in variation along the Elkhorn. Almost as soon as non-Indians arrived on its banks they began to tinker with the environment. Applying the standard proposed by Klein and Zellmer, certain stretches of the Elkhorn no longer qualify as natural. According to Steinberg's analysis, the damages inflicted by high water in the river cannot be considered natural disasters and the hardy pioneers not only carved a place for themselves on the Plains but also acted as architects of future destruction. What we have here, then, is a record of making, living with, and attempting to unmake unnatural disaster in a Great Plains river basin. This book argues that the settler colonists who came to the Elkhorn Basin in the nineteenth century planted a culture dependent upon a river it intended to tame, but could not, bequeathing to future generations a legacy of flood damage that cannot be attributed to "natural" disaster and has provided the focal point in the relationship between people and the river. It shows how struggles to alter rivers in the name of flood control played out not only in densely populated urban areas, but also on the sparsely populated Plains where they followed a different trajectory.

This book covers a lot of ground chronologically, with a heavy emphasis on the nineteenth and twentieth centuries, but does not cover much ground, or water, physically. Chapter 1 provides a natural history of the Elkhorn Basin, which illuminates characteristics that qualify it as a Great Plains river. The chapter also offers a primer on flood mechanics and a discussion of the historical significance of floods. Chapter 2 brings people into the Elkhorn Basin's story from the earliest records through the 1890s. This discussion extends from what we know about Plains Indians who lived in and near the Elkhorn Basin to the era of settler colonialism extending onto the Great Plains with newcomers figuring out how to co-exist with rivers that didn't quite live up to expectations set by rivers in other parts of the world. Chapter 3 narrates developments, many of them tinged by themes related to the Progressive Era, from the 1890s through the 1940s. This period saw the Elkhorn achieve arguably its greatest reach into the lives of settlers and their heirs, especially in and around

Neligh, but also the fracture of that tight relationship and an estrangement that left flooding the main feature linking people and the river. Chapter 4, the most tightly focused chronologically, tells of the Elkhorn's worst floods to date, the devastating crests of 1944, which brought significant intervention by the federal government in the basin. Chapter 5 follows developments inspired by the 1944 floods through the flood of 2010, which knocked 1944 from the throne of "worst ever." The conclusion offers thoughts on what lessons these stories might teach us and peers into possible futures with Great Plains rivers as climate change threatens to make already volatile central continental weather patterns even more so.

The Elkhorn, a commercially un-navigable, relatively inconsequential Great Plains river that flows through what many outsiders consider fly-over country and the people who have lived with it for decades have something to tell us. As in many relationships, things move along unremarkably much of the time. Occasionally, though, one partner in this relationship, the river, rages out of control. Or so the accounts by humans would have us believe. The people in this story might see in those rages evidence of sociopathy. But, really, the river just does what rivers do. Describing a flooding river as "raging" and "destructive" tells us more about the people than it does about the river. Studying how those people have used the river and its environs tells us even more about them. For its part, the river in this story might see in the human responses to it evidence of sociopathy. But as hard as it is for historians to interpret the human past, it is even harder to divine the non-human past. Whatever the Elkhorn might be thinking, we know what it keeps doing, which is rolling along … and flooding.

Chapter 1

THE ELKHORN RIVER BASIN

"Historical experience points to an indeterminate release to man of such 'new resources' as he becomes technologically capable of their utilization."

James Malin in *Ecology and History*[1]

"Floods exhibit to us a grand and astonishing spectacle. They swell the mind with the idea of awful and irresistible power. They shew us the vanity and impotence of man."

Joseph Lathrop, D.D., pastor of the First Church in West Springfield, Massachusetts, in a sermon inspired by a "succession of inundations," January 22, 1807[2]

About forty miles east of the middle of nowhere, groundwater flows to the surface of Nebraska's Sandhills giving birth to the Elkhorn River.[3] The Elkhorn is neither Nebraska's largest river nor its most important. Those distinctions would likely go to the Platte or the Missouri. But the Elkhorn starts and ends entirely within the Great Plains, making it distinctive, and its basin a laboratory for examining human experiences in the North American Plains as well as the region's environment. This chapter sketches the Elkhorn and provides a contextual discussion of flooding, a recurring and considerable problem.

Even within a single basin, the Elkhorn presents a story of two rivers. At its source and through much of its upper basin, sandy soils absorb much of the rain and snow that fall on the surrounding countryside. The river, itself, meanders sluggishly. In certain places it's difficult to figure out where exactly the channel is. As recently as the 1940s a practical-minded observer noted that

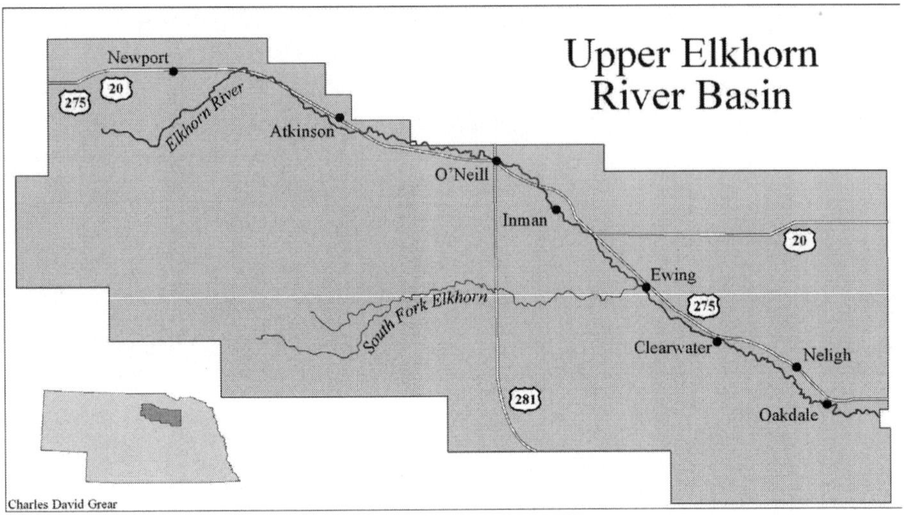

Map 1-1: The Upper Elkhorn River Basin. Map by Charles David Grear.

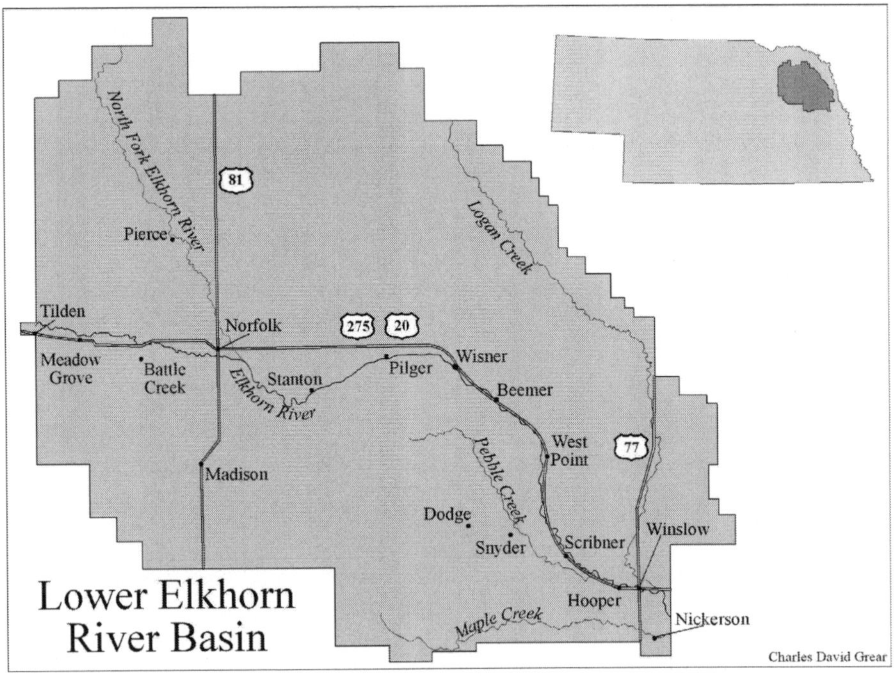

Map 1-2: The Lower Elkhorn River Basin. Map by Charles David Grear.

even thick underbrush could reroute the river in such stretches and send it into farm fields. Constructing a new channel and confining the river to a definite course could reclaim many acres of hay land. For the most part, flooding problems in the upper basin concentrate in the eastern reaches as the Elkhorn flows into its lower basin (Map 1-1).

Eventually, the Elkhorn becomes a "swift, clear, beautiful stream."[4] That swiftness reflects a change in topography as the landscape rises more precipitously from the river and the river falls more steeply as it seeks its mouth. The land itself changes, too, as the soils become less sandy and hold more clay and other constituents. These characteristics of the lower basin that make a swift, beautiful stream also create more runoff from precipitation. Beautiful as it might be in these stretches, the Elkhorn also becomes more prone to flooding and more likely to inspire descriptors such as destructive, rampaging, and mean.[5] Most of the action in the river's flood history took place in the lower basin.

From north-central Nebraska, the Elkhorn flows southeasterly to its mouth, which empties into the Platte just west of Nebraska's largest city, Omaha. Once the basin became part of the United States, humans imposed geopolitical boundaries upon the basin. A quick review of those units and present landmarks will help the reader form a mental image of the basin. Ignoring the land's natural contours, the newcomers sectioned off the land with right angles and straight lines to create the counties through which the Elkhorn flows. From west to east those counties and towns, noted parenthetically after their home county, include Brown, Rock, Holt (Atkinson, O'Neill, Inman, and Ewing), Antelope (Clearwater, Neligh, Oakdale, and Tilden, straddling the border of Antelope and Madison counties), Madison (Meadow Creek, Battle Creek, and Norfolk), Stanton (Stanton and Pilger), Cuming (Wisner, Beemer, and West Point), Washington (Arlington), Dodge (Crowell, Fremont—a Platte River city, yet near and important to the Elkhorn—Scribner, Hooper, Winslow, and Nickerson), Douglas (Waterloo and Elkhorn), and Sarpy (Gretna). Parts of twelve other counties also lie within the Elkhorn Basin. (See Map 1-2.)

A number of smaller streams feed the Elkhorn. Several play important roles in the pages that follow. Major tributaries include the South Fork (two of these exist, one flowing from Rock County into Holt, the other flowing entirely within Holt County); Dry Creek, also a Holt County stream; the North Fork, which flows south through Pierce County and past the city of Pierce;

Humbug Creek, which flows south from Wayne County into the Elkhorn at Pilger; Pebble Creek, which flows southeast through Cuming and into Dodge County where it empties into the river; Logan Creek, which drains the basin's northeast boundary and flows through Cedar, Dixon, Thurston, Cuming, Burt, and Dodge counties before joining the river at Winslow; and, Maple Creek, which flows east across Dodge County before meeting the river near Nickerson.

The Elkhorn drains a basin created by a glacier of the medial Pleistocene Age.[6] The river's gradual and consistent gradient lacks falls and notable dams. The remains of at least one mill dam, at Neligh, stood for decades until it washed out in the flood of 2010. Starting at an altitude of about 2,300 feet, the Elkhorn flows down its steepest gradient, more than 4.7 feet per mile, out of the northeastern Sandhills, a unique region in central and northern Nebraska roughly the size of Vermont and New Hampshire combined, that constitutes the largest sand dune area in the Western Hemisphere.[7] Sand dunes as high as 400 feet and as long as 20 miles give the Sandhills a desolate, otherworldly character painted in a mind-boggling variety of tans and browns broken by spring greenery and punctuated by wildflowers and brilliant sunsets. Many homesteaders cursed the gritty terrain and the winds that blow relentlessly across it, but the grasses holding the dunes in place make for some of the world's best cattle grazing country.[8] The chief crop in the western third of the basin is hay.[9]

As the river rolls through Holt County, its gradient eases to 4.7 feet per mile and less as it leaves the Sandhills and enters the fertile farming soils, finer soils laced with silt and clay, of the eastern basin. Here ranching gives way to agriculture dominated by corn cultivation and hog raising. In the middle basin, where Stanton and Cuming counties border, the basin's ranching and agricultural economies meet in symbiosis at feed lots where western basin cattle fatten on eastern basin corn and western basin hay. By the time the Elkhorn empties into the Platte, at an elevation of 1,074 feet, it has fallen slightly more than 1,200 feet—less than a quarter of a mile—on its journey through Nebraska's range and farmlands.[10]

All of the basin's soils developed in the grassland environment, which subsequently evolved. A tributary of the Platte, the Elkhorn drains about 7,000 square miles, all within Nebraska's borders. About 60 percent of the river's annual volume of nearly 1 million acre-feet (an acre-foot is enough water to cover an area of one acre with one foot of water) flows during March, April, May, and June—about 33 percent of the calendar year.[11] Put another way, the river's flow varies significantly during the course of a year. Generally well defined, basin boundaries extend from the Sandhills in the West to the Platte River Basin in the East. Although the Elkhorn's drainage area constitutes less than 8 percent of the Platte's total drainage area, the Elkhorn contributes significantly more to the Platte's total flow, as much as 22.6 percent as measured in the decade 1959–1968. This is due largely to the greater precipitation in the eastern basin and substantial irrigation diversions from the Platte in Colorado, Wyoming, and western Nebraska.[12]

Average annual precipitation varies from twenty-one inches in the western basin to twenty-nine inches in the eastern basin. Considered overall, the basin receives twenty-five and one-quarter inches of precipitation annually. The spring and summer months account for 75 percent of the basin's annual precipitation. June tends to be the wettest month, January the driest. Means, however, can be misleading because of the erratic nature of mid-continent weather patterns.[13]

The Elkhorn's sensitivity to flooding varies throughout the year and throughout the basin, depending on the soil's ability to absorb moisture. Floods in winter and early spring result from lesser amounts of precipitation than do floods in the spring, summer, and fall. Winter and early spring floods between 1959 and 1968, for example, occurred when the soil was either frozen or temporarily saturated and could not absorb water from a rapid thawing of a thick snow cover or from a light rain in combination with thawing. Widespread heavy rains caused the warm-weather floods. For three reasons, warm-season floods occur only as a result of excessive rainfall. First, vegetation takes moisture from the ground, increasing the soil's ability to absorb moisture. Second, vegetative cover slows the overland flow of water. And, third, warm temperatures boost evaporation rates.[14]

Left to its own devices, the Elkhorn wanders from one side of its floodplain to the other. A wandering river creates problems for people who rely on fixed property lines and permanent settlements, so since the late 1800s humans have in several areas tried to make the Elkhorn stay put by cutting a channel for it. Comparisons of the basin's earliest maps with recent aerial photographs show many channel changes over the years. Oxbow lakes and other abandoned channels bear mute testimony to the Elkhorn's wandering ways and human attempts to reform them.

Several factors combine to allow the Elkhorn to forge itself new channels. As mentioned, it carries a heavy sediment load and when its current slows as it exits the Sandhills, the river dumps silt into its channel. Over time the channel fills in and the Elkhorn seeks lower ground and a new channel. In that respect, the Elkhorn fits Walter Prescott Webb's model of a typical Great Plains river.[15] Also, lack of exposed bedrock along the channel allows the Elkhorn to roam. The river flows over a malleable sand bed for its entire length. So readily does the sand bed adapt to flow changes that the channel's deepest point one day may be replaced by a sandbar the next. The Elkhorn's flow varies with a volatility to match the channel's. Recorded peak flows have ranged from 40 to 108,000 times the recorded minimum flows. Some of the most concentrated thunderstorms in the United States occur in eastern Nebraska, contributing to wide variances in the Elkhorn's flow.[16] In combination, the river's sediment load and decelerating flow, its sandy bed, and drastic changes in flow make the Elkhorn a mercurial, mobile river.[17] Despite an absence of falls, the Elkhorn's shallowness and highly variable flow render commercial navigation impossible. People have tried, with varying degrees of success, to put the river to work in other ways. In at least twenty-four places they harnessed the river, mostly for small milling operations, virtually all of which have been abandoned for decades. Even at the best sites, the "generally erratic flow precludes practical development."[18] Erratic. Inclined to wander. Volatile. This is not a river that is easy to live with even under normal conditions.

Throughout the Sandhills, the Elkhorn moves at its most rapid pace and carries its largest load of sand. Sandy soils there allow much of the western basin's precipitation to sink into the ground and keeps overland runoff relatively low.[19] Such soils do not, however, hold water well against gravity. Thus they tend to be arid unless mixed with loam or very near the water table. But because the Sandhills' vegetation, a unique community of grasses, sedges, and forbs, remains relatively undisturbed and because the area sits atop a large groundwater reservoir, it has been less susceptible to the effects of drought than have other regions of Nebraska.[20] In spite of its aridity, the western basin's soils produce above average crop yields when planted in crops adapted to the soil, moisture, and temperature conditions making it excellent grazing and hay country.[21]

Vegetation also helps explain why the upper Elkhorn is less flood prone than the lower Elkhorn. From mid-February to the end of October, the Sandhills environment needs more moisture than precipitation provides. The system's need for water and the region's porous soil absorb spring and summer rains like a sponge.[22] Entering the eastern basin's farmland, where finer, more fertile soils hold water longer than the west's sandy soils, the Elkhorn slows and undercuts its banks.[23]

Eastern basin soils absorb moisture more slowly, but can retain much more than the western basin's sandy soils. Here, also, the valley rises more steeply from the river, 200 feet per mile in the eastern third of the basin. The steeper slope, combined with a soil that absorbs moisture slowly, translates into rapid overland runoff in the eastern basin, which can be considerable during periods of intense rainfall or rapidly melting snow. But the eastern basin's moister, richer soils are well-suited to agriculture and constitute the basin's best tillable land.[24]

Plants rooted in the Elkhorn Basin's fertile soils facilitate drainage. Through evapotranspiration, the process by which plants draw water from the ground into their roots and release it into the atmosphere through their leaves, Elkhorn Basin flora discharge a great quantity of groundwater. The upper basin's extensive hay lands, timber and brush bordering the river downstream from Ewing, and sub-irrigated hay, alfalfa, and other bottomland vegetation in the lower basin all play a role.[25] Because streamflow during months of low precipitation consists entirely, or nearly entirely, of groundwater effluent plus overland runoff, vegetation's effects on the Elkhorn's flow can be gauged.[26] Generally, dry-weather streamflow when vegetation is dormant far exceeds dry-weather streamflow during the growing season—sometimes by more than twice the volume.[27]

In addition to grasses and agricultural flora, forest covers about 3.5 percent of the basin. Most of the timber grows on the banks of the Elkhorn and its tributaries or in planted woodlots or shelterbelts. Naturally occurring trees include elm, ash, cottonwood, and oak. Settlers also planted some evergreens, mainly Scotch and Austrian pines, for use as Christmas trees.[28]

The Elkhorn and its tributary streams provide a home for thirty-five major fish species, including largemouth bass, channel catfish, flathead catfish, northern pike, white crappie, black bullhead, and carp. The bluntnose minnow, classified a rare species in Nebraska, also lives in the basin's streams. At one time the basin's ninety-three streams provided 1,550 miles of fish habitat. Channelization projects, however, eliminated 291 stream miles (18.8 percent) to the detriment of fish habitat.[29]

A variety of wildlife lives in the basin's riparian lands. Scarce to moderate densities of white-tailed and mule deer live throughout the basin as do low to moderate densities of antelope in the Sandhills region. Pheasant, bobwhite quail, prairie chickens, and sharp-tailed grouse are among the game species present in various portions of the basin. Cotton-tail rabbits and fox squirrels are common where habitat allows. And, as one would expect, waterfowl grace the region in large numbers and variety. Mallards, blue-winged teal, and gadwall are among the most common nesting species.[30]

Flood Mechanics

Almost always the Elkhorn flows unremarkably within a typical range. But, like a great many rivers, the Elkhorn occasionally carries exceptionally large amounts of water and expands beyond its typical limits. This has happened for a very long time and while it cannot be described as regular or predictable, it can be considered normal and part of the Elkhorn's nature. Humans call these events floods and tend to categorize them as natural disasters, a label that reveals much about perspective and historical development.

Floods seem deceptively simple at first glimpse. A high volume of water enters a stream and the stream overflows its banks. A number of different events can cause flooding of various types depending on the stream in question. Floods, too, have different consequences depending on what organisms or systems are examined. The human point of view occupies the chief concern here, but it should be noted that floods exert significant impact upon other animals, plants, and the landscape of the flooded stream valley. Floods can also seem deceptively similar—river rises, does a little or a lot of damage, river recedes. But individual flood events can be quite dissimilar even in the same basin and in the same place within a basin. Water might rise quickly when a thunderstorm dumps heavy rains on saturated soils resulting in massive runoff into a river. Such a flood might allow little time to evacuate or to move livestock to high ground resulting in shocking, traumatic losses. Water might rise slowly and steadily thanks to a main-channel ice jam that impedes drainage of snowmelt. Such a flood might allow evacuation or sandbagging or other defensive measures and might be annoying, but not necessarily shocking or traumatic.[31]

Much of the drama in this story plays on a stage known as floodplains, the lowlands bordering rivers. Since rivers are unable to create channels that would never overflow, the floodplain is actually part of the channel during times of heavy flow.[32] Each river builds its own floodplain by dumping its load of sediment and continually reworking it through erosion and subsequent flooding. This invites human usage, since the river normally remains in its channel leaving the fertile soil of the floodplain dry. Early civilizations developed in these areas, which continue to serve as important farmlands, yet the floodplain unmistakably belongs to the river.

A typical river tends to occupy only a portion of the valley trough which it meanders. As it flows, the river's water bounces off the harder alluvial deposits of its channel seeking the path of least resistance over and through softer deposits. At bends, however, the river's greater velocity on its outside edge erodes that bank and allows sediment to be deposited along its inside bank where velocity is lower. As this process continues the river moves through its valley. Thus the floodplain is the earthen portion of the dynamic riverine system.

All over the world humans have settled on these "living" floodplains for very good reasons. It puts them near a source of water and allows them to make use of the typically fertile soil. In some cases, including several Elkhorn settlements, a river's current as a source of power for milling beckoned humans to inhabit the floodplain. According to William G. Hoyt and Walter B. Langbein of the US Geological Survey, the relation between floods and topography as an aspect of human geography has been neglected by scholars. It seems plain, however, that settlement patterns in the Elkhorn's basin, with towns built around mills, which necessarily had to be built near running water during a certain era, led to enormous potential for flood damage as buildings, businesses, and homes sprang up on the floodplain.[33] The next chapter explores this theme in detail.

Typically, a river system's most vivid actions are floods. Simply put, a flood is a flow in excess of channel capacity and is a natural river characteristic.[34] According to Hoyt and Langbein, "Floods are as much a part of the phenomena of the landscape as are hills and valleys; they are natural features to be lived with, features which require certain adjustments on our part."[35] Several phenomena cause floods. Hurricanes can cause torrential downpours and flooding but are most influential in coastal areas. Rainstorms, the mechanism by which "transient rivers of vapor in the blanket of air enfolding the earth" feed the rivers of water flowing over the earth's surface, can also cause streams to overflow. Rainstorms occur when a moisture-laden air mass cools, the moisture condenses and falls to the earth as rain.[36] Thunderstorms, spectacular rainstorms accompanied by lightning, thunder, gusty winds, and sometimes hail, account for most floods on small watersheds. Short-lived and concentrated, rarely covering more than 200 square miles, the thunderstorm occurs as overheated, moist air rises rapidly then cools, releasing its moisture to the accompaniment of thunder and lightning.[37]

Nebraska and the Elkhorn Basin lie within the approximately two-thirds of the United States where cold temperatures allow the accumulation of snow and where either snow or ice can play a major role in flooding. Melted snow, however, rarely acts alone in causing flooding. But in tandem with a rainstorm, the effects can be devastating. A rainstorm not only adds more water to an impending flood, but normally arrives on the wings of a warm, moist wind. The wind melts the snow while the moisture condenses on the snow, producing additional heat for melting (0.1 inch of condensation melts 0.7 inches of snow water). Remarkably, the rainwater itself performs little melting. (At fifty-two degrees, one inch of rain produces only one-seventh inch of meltwater.) In fact, a heavy snow cover can absorb significant amounts of rain to be released during later melts.

Snow melts from the surface down and from the bottom up. Early in the melting process, meltwater freezes as it seeps down through the pack. As the pack becomes coarser, its density increases and its sponge-like ability to hold water decreases. By the time the melting zone reaches the ground surface, the rate of melting is rapid and the pack is generally saturated, or "ripe," with meltwater. When most of the snow has become slush, subsequent heat breaks down the snow, allowing a relatively quick release of water into surrounding streams. Flooding, or lack thereof, depends on the amount of accumulated snow and the type of and intensity of the thaw. The combination most likely to produce flooding is a snowy winter followed by a rainy thaw.[38]

A sudden thaw after a long, hard winter can also bring flood danger from ice formed within a stream. A stream swelling with rain or meltwater can crack its ice cover, setting pieces floating downstream. When chunks of the broken ice cover are thick enough, they can lodge in narrow parts of the channel, in bends, or at bridge supports where they form an icy dam. The size of the dam and the amount of restricted water increases until the pressure or thaw becomes great enough to break it. It is possible that the ice dam can cause flooding behind the dam. The sudden release of ice and water can then threaten to flood downstream areas.[39]

Flooding in the Wide World

Although scientists now describe floods as a natural part of a river's being, the sudden rise of a normally docile, serene river out of its banks and onto its floodplain, where it destroys human structures and sucks humans and animals to their deaths in its swirling dark waters, has made floods seem evil and malicious to people for centuries. When accompanied by the crashing breakup of ice or the spectacular electrical show of a thunderstorm, floods can indeed be terrifying events. Peoples around the world have traditions about the near eradication of humans, save a very few, by natural disaster. The Arawaks of Guyana tell of a tremendous fire. Zoroastrians of ancient Persia believed a great freeze depopulated the earth. Perhaps the most common belief, however, is that a flood nearly cleansed the earth of humans. One collection of such flood tales boasts 302 texts from around the world, including Mesopotamia, Egypt, Greece, Syro-Palestine, Europe, Indian, eastern Asia, New Guinea, Central America, North America, Melanesia, Polynesia, Micronesia, Australia, and South America.[40] Best known in the Judeo-Christian tradition is the biblical tale of Noah given in the book of Genesis. After creating humans, the Lord becomes displeased with their wickedness and decides to destroy them with a flood. Noah, however, had pleased the Lord, who told Noah how to save his family and the earth's animals by building a special boat. Safely on the boat, Noah's people and the animals survived a rainstorm of forty days and forty nights, which caused not one, but all, rivers to overflow their channels and floodwaters to cover even the tops of mountains before receding.[41] Being linked to the near extermination of humanity, it is little wonder that floods have had a bad reputation for thousands of years among people of diverse and dispersed cultures.

In other cases, floods have benefitted humans or at least struck a tolerant balance between destruction and benefit. Egyptians depended upon the Nile River to flood its valley with fertile soil and for moisture to bring a good harvest. Bumper crops fed the people and allowed the government to tax with relative immunity from unrest. In the Far East, floods have created fertile plains and deltas upon which millions of people live and farm, in spite of a constant threat of flooding. In China, for instance, millions live densely packed on floodplains, which have been inundated since well before the country's first recorded flood, the Yellow River flood of 2297 BCE. The Yellow River's nickname, "China's Sorrow," reflects its flood-related infamy. Even though thousands of Chinese drown annually and a single flood can affect millions of people, moving out of the floodplain would mean starvation.[42]

Experiences of British colonists in Australia in the late eighteenth century and early nineteenth century present similar issues, but, perhaps, ones more relevant to considering the Elkhorn's past as both cases involved settler colonists of European extraction moving into new landscapes. When the British moved into New South Wales, they encountered not only a new land, but also new rivers. Australian rivers differed from British rivers in hydrology, ecology, and soil composition. They had different flow rates. Of special note, flow rates of Australian rivers could be as much as twice as variable as rivers in other parts of the world. In other words, they were not very stable and have, in fact, been described as "non-equilibrium" rivers. British newcomers, who brought with them a colonial model based on building farms and settlements in stable landscapes and climates, found themselves trying to execute that model in dynamic landscapes, "zones of flux."[43]

Whatever colonists noticed about the *nature* of Australian rivers, at least some of them recognized signs of flooding on the landscape, noted the dangers of floodplains, and avoided and discouraged settlement in the floodplains. The governor who implemented the first stages of settlement avoided granting land along the Hawkesbury River in New South Wales partly because of flood danger. He also worried about governing a settlement far removed from the seat of power in Sydney. Despite that official guidance, settlers on the remote frontier independently began taking riparian lands from aboriginal peoples in late 1793 or early 1794. Within months, in March 1794, a flood hit the settlements. Two more floods hit in 1795. In 1799 the largest flood recorded to date, more than double the height of the bigger of two 1795 floods, hit the settlement. Despite immediate and repeated flooding, most settlers decided to stay.[44]

In 1810, less than twenty years after the first settlements in the Hawkesbury-Nepean valley, the region's governor responded to the record of devastating floods by selecting sites for five flood-proof towns and opening a campaign to relocate established settlers to them. The plan's apparent logic did little to persuade most settlers, who chose to remain on their flood-prone claims. Between 1811 and 1817 the governor tried and then tried harder to effect the resettlement, still with little success. He ended the previous policy of providing no-cost government flood assistance to victims. Instead, flood victims would have to repay the government for any aid. When that didn't work, he threatened that those who remained in the floodplain would receive no government aid at all in event of flood. In effect, this was an early declaration that any flood damage from that point on would not be a natural disaster, but the fault of settlers who refused to move out of harm's way. Nothing worked.[45]

Trying to make sense of the settlers' choices, the governor concluded they suffered from an "Infatuated Obstinacy," that they had been, as historian Grace Karskens put it, "bewitched by their land." The governor's own reports, though, show he understood quite a lot about the nature of that bewitchment:

"[T]he lands on the banks of the Hawkesbury are so very productive and Cultivated with such very little Trouble that it is not at all Surprising that the Old Settlers who have so long lived on these Farms should be so partial to, and so unwilling to leave them."[46]

The governor viewed floods from a perspective that considered moving to higher ground as a rational response. The settlers, though, saw them differently. Floods were unpredictable, wild forces of nature that brought risks of death and destruction considered a fact of living with a river. The sea could drown them, but they had sailed it to arrive in Australia. Similarly, disease, accidents, even death could strike at any time and might stem from luck, fate, or an act of God. And floods brought not just death and destruction, but the reward of rich earth.[47] In the middle of North America, calculations about living with rivers and dealing with their floods followed similar contours.

Floods in the Great Plains

A sketch of flooding in the Missouri River Basin, the Great Plains' largest, will help set the stage for the intensive study of the Elkhorn in the chapters that follow. Records of flooding are scarce for the period before settler colonists of European extraction showed up. To get a sense of what flooding meant for Plains Indians, we can look to experiences of tribes such as the Dakotas, Lakotas, Mandans, and Hidatsas who lived on the banks of the Missouri and its tributaries. As with other people elsewhere, Plains Indians lived with rivers, using them for various purposes and deriving benefits from the riparian environment. They also died on occasion and suffered other consequences when rivers overflowed.

One of the earliest recorded floods appears on a Lakota winter count, a type of historical record using pictographs, often drawn on animal hide, to record a notable event for each year that passed. One such count identifies 1711–1712 as "Four lodges drowned winter." It shows water, about waist-high, enveloping four tipis, or lodges. Evidently, some Dakotas who had sought shelter from the Plains wind in a bottomland campsite in the spring of 1712 got caught in a flash flood when an unexpectedly early spring thunderstorm brought a nearby creek rapidly out of its banks. High water swept away four lodges and drowned the occupants.[48]

On April 6, 1826, the Missouri rose a whopping seventeen feet in a matter of hours causing great trouble for indigenous peoples living near it. Mandans who lived at Mih-tutta-hang-kusch, which sat on a bluff, watched the rise with awe, but from a vantage point that turned out to be safely above even this remarkable rise. Another Mandan town just upstream and two Hidatsa towns didn't fare as well—the Missouri covered them.[49]

Multiple Lakota winter counts recorded the horrific Missouri River 1826 flood. Broken ice brought the Missouri suddenly and unexpectedly out of its banks just below the Whetstone River. Dakotas, still sheltering from winter winds in the bottoms, scrambled onto ice floes or climbed trees to escape the frigid water. Others were crushed by chunks of ice and many were swept away or drowned. One winter count estimated the loss at thirty lodges and another noted the deaths of women and children. Records don't mention property losses, but the destruction of thirty lodges suggests that the group lost not only shelter but also a considerable amount of food, clothing, and other goods.[50]

Indigenous peoples who lived near rivers expected and planned for seasonal high water and also took advantage of occasional food windfalls provided by waterways. They typically located their biggest and more permanent settlements on high ground. They benefited from the rich alluvial soils, which they farmed. They also put the river to work as they procured meat, sometimes using the Missouri effectively in winter bison hunts. Groups of Mandans would drive one or preferably many bison onto the frozen river and then toward thin spots in the ice. When the animals' weight sent them crashing through the ice, the current swept them under remaining ice where they drowned. Other Mandans waited to collect the carcasses at open water downstream. A skilled and fortunate band of Dakotas procured a year's worth of meat using a similar tactic during the winter of 1702–1703. The Dakotas, though, were able to use the river as cold storage when it refroze around the thousand or so head that had fallen through. For months, when hunger struck they were able to go to the river and cut from the ice as many bison as they wanted to eat. On at least one occasion, they were able to get a winter's supply of fish the same way. Missouri River peoples, and presumably those of the Elkhorn as well, considered "float bison" a great treat. As these stories attest, river bottoms, sheltered from the biting wind, provided a fine place for winter camps. Their fertile soils made good farmland and rivers could provide bounteous, if erratic, supplies of fish and meat. On the other hand, living in the bottoms and with rivers caused inconvenience and, in the case of a flood such as that of 1826, destruction of homes, property, foodstuffs stored in underground cache pits, and even death.[51]

In 1881, ice jams and melting snow created a flood that damaged recently established non-Indian settler colonies in the Dakota Territory and Nebraska. The winter of 1880–1881 hit early and hard. Blizzards struck lower Dakota in mid-October and were followed by a hard freeze, which trapped steamboats in deep ice on the Missouri and froze its tributaries as well. Heavy snows in January blocked wagon roads and railroads while continued cold created ice thirty-eight inches thick on parts of the Missouri. By March 19, two feet of snow covered the ground in parts of Dakota. A thaw on the Missouri's upper reaches broke the ice there releasing ice and snowmelt that flooded portions of Montana. Cold weather in Dakota kept the Missouri frozen and allowed more snow to accumulate. At the freeze line on the river, an ice dam formed and caused flooding behind it. In places, runoff flowed over the ice on its way downstream. When the ice broke at Bismarck, it formed a gorge nearby at Fort Abraham Lincoln and sent floodwaters into Mandan. East Pierre (now known as Pierre) felt the flood on March 27 when waters rose four to six feet in one day. Eight days later another rise soaked the town. Further downstream, at Niobrara, Nebraska, separate rises occurred on March 30 and April 6. Each brought three to five feet of water into the town and sent its inhabitants scurrying to higher ground. When the long, hard winter of 1880–1881 ended, it brought further hardship to those humans who had endured it.[52]

Despite relatively light settlement in Dakota and Montana, the winter floods caused considerable damage. Reports from Montana said Indians (not identified by tribe) camped along the Missouri suffered eight deaths, lost 500 or more horses to drowning, and abandoned their camp so hastily that they left behind tepees, robes, and most all their possessions. Non-Indians and their livestock perished, too, in the icy inundation. Ice three to six feet thick buried the streets of Mandan, but since flooding and ice hampered transportation into the town, merchants had only small supplies of goods on hand and thereby suffered limited losses. One railroad lost a reported 20,000 bridge timbers, but had hopes of recovering some. Another, the Dakota Southern, was nearly destroyed from Sioux City to Yankton. High water left the streets of Niobrara full of mud and debris but swept away all one- and two-story frame houses in Green Island, Nebraska. Dakota Territory suffered a loss of several thousand acres when the Missouri cut a new channel during the flood, shortening its route by seventeen miles in the process. Vermillion, Dakota Territory, once a riverfront town, found itself four miles inland and relieved of forty buildings swept away by water or crushed by ice. In dollars, flood loss estimates totaled about $1 million, of which $200,000 came from Vermillion.[53]

Human responses to the Missouri River flood of 1881 seem almost comical, considered from the safe distance of more than a century. The first reaction was to run away. But there was nothing funny about Indians camped on the Missouri being awakened by a surge of water, which sent them to high ground in a "race with death." Citizens of East Pierre organized a squad of vigilantes to guard against looters. Other locals responded to the flood by looting. The vigilantes caught them and ejected them from town. New York City's Produce Exchange responded by soliciting donations for victims and the territorial governor, Nehemiah G. Ordway, in Washington during the fateful winter, tried to raise $100,000 for relief aid. The secretary of war ordered the military to dispense supplies from government stores along the river from Yankton to Vermillion. The towns of Niobrara and Vermillion responded by relocating to higher ground. Moving to high ground would separate the village from Missouri River steamboat traffic and devalue land in the old town. Niobrara's residents decisively implemented their response—escape—within weeks of the flood as teamsters hauled homes and businesses a mile and a half south and west of the old town site. The 150 residents of Green Island, though, let their town die, deciding to relocate rather than rebuild.[54]

Missouri River Valley inhabitants seemed wise to abandon the floodplain in Vermillion, Niobrara, and Green Island. People who experienced the Missouri River flood were surprised by the sudden nature of the 1881 flood. They also, in most cases, did not have a deeply vested interest in remaining in the floodplain. Founders had invested only about one human generation in developing these relatively new communities (Niobrara was about twenty-four years old in 1881), and it made sense to find a better location than a floodplain if people were going to remain there.

Another series of Missouri River floods deserves mention here for the response they provoked. Beginning on April 12, 1943, the Missouri rose from Pierre, South Dakota, to Rulo, Nebraska, reaching heights not seen since 1881. The flood, attributed to rapidly melting snow on the northern Great Plains, covered about seven hundred thousand acres in the upper Missouri Basin and an estimated $7.8 million in direct and indirect damages. Much of Omaha was under water and its airport—important to national defense—was entirely dislocated. The lower Missouri Basin suffered from flooding induced by heavy rains from May 6 to 11 and from May 15 to 20. The flood covered 540,000 acres, caused $7 million in damage, and killed three people. Finally, the middle Missouri Basin flooded in mid-June when storms from June 8 to 16 dumped six to seven inches of rain on southeast Nebraska, southwest Iowa, northwest Missouri, and north central and northeast Kansas. The flood submerged about 960,000 acres in the area for the first time that year and soaked another 280,000 acres that had been hit by the April inundation.[55]

These flood emergencies prompted the federal government to respond with legislation in 1944 and 1945 that created a development plan, which came to be known as the Pick-Sloan Plan, for the Missouri Basin. The plan called for immense construction projects for purposes including flood control, irrigation, power generation, and navigation.[56] Most important for this discussion, it focused the federal government's attention on the Missouri Basin, which includes the Elkhorn Basin, in 1944 as Elkhorn floods caused unprecedented damage. As will be discussed in a later chapter, it not only paved the way for federal intervention in the Elkhorn Basin, but also influenced the nature of that intervention.

This contextual survey could go on, but doing that would not substantially change the outline established thus far. Rivers provide a number of benefits to humans, including water, power, fertile soil, and transportation. But as naturally as they flow, rivers overflow and produce floods. This natural action is how rivers deposit the fertile soil so coveted by humans in their floodplains. Humans have utilized the Elkhorn's floodplain soils for crop production for centuries. But for most of that time they built their dwellings above the floodplain, keeping their valuables safe from floodwaters. The most recent arrivals in the valley, however, came equipped with technology that allowed them to harness the Elkhorn's power for their economic benefit. Settlers of European extraction continued to farm the fertile floodplain, but they also placed their mills, to convert their crops into commercial goods, in the floodplain to harness the river's power. The ability of mills to convert crops into a commercial product made them hubs of commercial activity. Often a town developed near a mill, creating more commercial opportunities. Towns, in turn, begged to be connected by transportation networks to facilitate access to markets. Thus the Elkhorn Basin's towns and transportation networks—the largest concentrations of capital in the valley—developed in the floodplains.

The settlers' desire to be near, and to reap the benefits of, rivers presented a problem: proximity bred not only profits, but also danger and destruction. As technology brought permanent residents into the Elkhorn's floodplains, it gave them false confidence that they could control the river and protect themselves and their possessions from flooding. A pattern emerged of intense or especially damaging flooding followed by human attempts to eliminate or control flooding to protect assets placed in the floodplain.

The Elkhorn Basin, a Great Plains sub-region dominated by non-Indian settler colonists beginning in the 1850s, offers a glimpse into more than a century-and-a-half of dialogue between human society and its environment as viewed through flooding and responses to it. How did humans respond? How much time passed from the moment a damaging flood occurred to the human response? Did valley inhabitants respond offensively, trying to control the river and its natural flooding, or did they respond defensively, accepting the inevitable floods and modifying the environment to coexist with the river? The history of major flooding in the Elkhorn Basin that follows will include a brief discussion of each flood's causes, a survey of damages, and an analysis of human responses to the floods and their significance to the Basin's environment. These flood stories reveal the character of life in the floodplain and reflect human faith in technology's ability to control nature.

Chapter 2

UP THE ELKHORN

"Building in a floodplain is like pitching your tent on a highway when there are no cars coming."

—Association of State Floodplain Managers[1]

A German duke, of all people, penned what is probably the first written record of an Elkhorn flood. Paul Wilhelm, Duke of Württemberg, who visited North America from 1822 to 1824, traveled up the Missouri River as far as present South Dakota in 1823. That trip took him overland through the Elkhorn Valley, where he experienced one of the violent storms characteristic of the Great Plains. On August 12, after covering about thirty miles in sweltering 102-degree heat, the duke's entourage camped near the Elkhorn. A "hurricane-like storm" blew from the southwest for several hours before the winds subsided, giving way to a torrential downpour. The rains crushed a shelter erected by the party and the wind blew down the duke's tent. The storm's runoff swept away the duke's tent and everything else not heavy enough to stay put.

Daylight revealed massive cottonwoods lying on the ground. The storm had blown down some and lightning had split others. Runoff had transformed the Elkhorn into a "torrent," which had flooded Indian earth lodges abandoned for the summer. The duke learned, to his surprise, that the crops growing in the Indians' fields survived the storm. Although he gave no basis for his conclusion, the duke reported that "the Indians are so clever in the choice of their various fields and so accustomed to such storms that the crops will rarely be harmed."[2] These unidentified Indians, then, had developed agricultural practices that coped with violent Great Plains weather patterns including flooding. This early written record of an Elkhorn flood, left to posterity by a tourist, suggests a great deal about living with the Elkhorn and it points to an historical watershed.

People have lived in Nebraska's fertile river valleys for thousands of years. Most of them clustered along rivers larger than the Elkhorn, namely the Missouri and the Platte, but from time to time people called the Elkhorn Valley, often by other names, home. We know little about how those people dealt with floods, but it is safe to assert that the river flowed out of its banks, destroyed crops planted in the bottomlands, and otherwise affected human activity along the banks ever since such activity began. The duke's story suggests that American Indians coped well with flooding. In that instance, the Indians weren't around to be chased from their homes by floodwaters. The seasonal cycle of Indian life mitigated flood inconvenience during part of the year. Furthermore, by luck or by design or both, the crops in the floodplain survived this particular flood. But the duke happened along just before big changes came to the basin. When non-Indians took up permanent residence in the Elkhorn Valley in the second half of the nineteenth century, things would change. Those people would treat the floodplain differently and that meant they would live with the river differently. They put buildings and roads, infrastructure intended to be permanent, in the floodplain. They counted on that infrastructure to provide vital connections to the outside world. Consequently, floods became more inconvenient and costly to humans, who began to fight the river and its floods to protect their property.

The Elkhorn Basin as Indian Country

Human activity in the Elkhorn Basin dates to approximately ten thousand years ago when the dialogue between humans and the basin environment began. Because no comprehensive archeological or anthropological study of human activity in the basin exists, studies of similar people from nearby regions supply information for this section. Spear points dated 8500 BCE once hurled by people of the Paleo-Indian tradition constitute the first record of human activity in the basin. Aside from the fact that these mammoth and bison hunters were present, we know little about them.[3]

Peoples of the Archaic tradition succeeded Paleo-Indians. Archaic peoples were active in the basin from about 8000 BCE to 1200 BCE. The technology and material culture of this age provided the basis for the Indian culture encountered by European explorers and settlers in and after 1500. Diversifying their economy from that of the meat-dependent Paleo-Indians, Archaic peoples used developing technology to exploit more of their environment through intensive hunting, gathering, and fishing. They developed specialized tools and utensils made of stone, bone, horn, ivory, shell, wood, hide, fiber, and copper, and invented techniques for preserving and storing food. The latter innovations, especially, allowed Archaic peoples to travel over a smaller area than their Paleo-Indian ancestors. But food still dictated their movements as they moved in small groups from one food source to another as the seasons changed.[4]

Since few Archaic artifacts have been recovered in the Elkhorn Basin, most of these people's activities remain mysterious; but, enough evidence left by the next cultural tradition, Woodland, active until about 1000 CE, exists to permit some speculation about their lifestyle. Woodland peoples typically lived in thatched or hide-covered huts on or near the Elkhorn's floodplain where they grew corn, squash, and other crops. They also hunted and were the first people there to employ bow-and-arrow technology and to use ceramics.[5]

People of the Nebraska Phase of the Central Plains Tradition succeeded the Woodland peoples in the basin, residing along portions of the Elkhorn's final fifteen miles. While the Nebraska Phase spans the years 1050 to 1425 CE, the most intensive activity along the Elkhorn probably occurred from 1250 to 1425 CE.[6] These peoples lived in villages of square earth lodges typically located on terraces or ridges ten to fifteen feet above the floodplain, although evidence exists of scattered lodges, perhaps winter quarters, located on the floodplain itself.[7] Single homes or groups of two to four homes were built with anywhere from a few yards to several hundred feet separating them. Storage pits, usually located under the floor, held food supplies and were perhaps a primary reason for building the lodges above the floodplain.[8] Nebraska Phase peoples limited their use of the floodplain to farming, hunting, and fishing.[9] Because researchers have found no European trade goods in Nebraska Phase sites, they speculate that the villages were abandoned before 1500 CE. Anthropologists offer reasons for abandonment ranging from invasions by competing indigenous nations to a drought, which made agriculture difficult and probably drove bison herds to better-watered locales.[10]

Sometime before 1600 CE, people of the Oneota culture reestablished a human presence in the region. Oneota peoples sometimes lived in earth lodges, but more often lived in less substantial homes of pole frames covered with mats. Dense clusters of remains indicate that these people lived in fairly permanent villages. For food, they grew corn, beans, and other crops, which they supplemented by hunting and gathering. Woodland animals played a more important role in Oneota life than did bison, and the people also fished streams near their villages. Excavations of Oneota sites, including one near the present town of Stanton, show that these people occupied the same sites as tribes that later lived in the area. Overlapping remains at these sites seem to indicate that Oneota people had contact with later tribes—contacted when non-Indian settlers came to the area—but make it difficult to determine the timing of such contact.[11]

One of those later tribes, the tribe most closely associated with the basin since non-Indian settlement, is the Omahas, whose name means "against the current" or "upstream," deriving from the tribe's historic movements up the Mississippi and Missouri rivers.[12] Once settled in Nebraska, the Omahas built their villages in an area bounded in the east by the Missouri River and in the west by the Elkhorn, known to the Omahas as "Waté."[13] A semi-sedentary tribe with a dual buffalo–agricultural economy, the Omahas lived in river or stream valleys, which provided them with fertile, easily worked, well-watered bottomland for their crops, timber to construct their lodges, flood-free and defensible terraces on which to build their villages, drinking water, and a source of game.[14] Beyond the Elkhorn lay Pawnee lands and the site of several battles between the two tribes.[15] Other tribes, including the Oto, Brulé and Oglala Lakota, and Cheyenne hunted in the Elkhorn Basin.[16]

During his travels on the Great Plains in 1823, Paul Wilhelm, Duke of Württemberg, visited a "large village of the Omahas on the Elkhorn"—one of the earliest accounts of the tribe living on the Elkhorn.[17] By 1830, if not earlier, advances by the Brulé Lakotas had driven the Omahas southwest from their previous village on Omaha Creek near present Homer, Dakota County, Nebraska, to the vicinity of the Elkhorn.[18] By the time non-Indian settlers began arriving in the Elkhorn region, they associated it most closely with the Omahas. As the duke's account at the beginning of this chapter suggests, the Omahas lived well with the Elkhorn. Their crops survived the thunderstorm he witnessed, but he noted that at least some of their dwellings felt the touch the Elkhorn's high water. As those dwellings appeared to have been seasonally vacant, it seems unlikely that their builders would have considered themselves victims of a natural disaster. In fact, it would appear that the Elkhorn and Omahas had established a peaceful coexistence on the floodplain, if not a mutual understanding of how to use it. That state of affairs was about to change.

Newcomers Reimagine the Elkhorn

Lewis and Clark beat Paul Wilhelm to the Elkhorn by almost twenty years, but they didn't write much about it. During their trek up the Missouri, the party briefly explored the Platte and in so doing came across a river Clark described as "about 100 yards wide with Clear water and a gravely Channel."[19] It was the Elkhorn. The party camped several days in the Elkhorn's vicinity, noting a large population of wolves as well as turkey, deer, grouse, and absolutely no flooding.[20]

Non-Indian travel through the central plains picked up notably with the exodus of the Church of Jesus Christ of Latter-day Saints from Illinois in 1846. These travelers, too, found that the Elkhorn could present a significant obstacle. Wishing to avoid contact with other overland migrants who might be hostile to their religion, the Saints, commonly known as Mormons, established the Mormon Trail to the valley of the Great Salt Lake on the north bank of the Platte River. That would segregate them from traffic on the Oregon Trail, which ran along the Platte's south bank. The Mormon Trail, then, crossed the Elkhorn near its mouth. Depending on streamflow, the Mormons encountered a river crossing that ranged from inconvenient to impassable.

Heavy rain and high water on the Elkhorn brought the Almon W. Babbitt Company of Mormon migrants to a standstill in the spring of 1851. Late in May a normally easy ford at the Elkhorn became a "roaring torrent." The Babbitt Company waited a week for the river to become fordable again. Once across, the company made only a mile through "country covered with water as far as we could see," Sarah Harris recalled. Again the company pitched camp and waited for conditions to improve. The high water flushed rattlesnakes from their homes and they, too, headed for the dry ground occupied by the Mormons. One woman found a rattlesnake crawling up her wagon's wheel as she dressed one morning. Another was found coiled under a carriage as children played nearby. The snakes evidently kept their fangs to themselves, but the same could not be said of the Elkhorn. It rose to axle or even wagon box height where it damaged animal feed and other property in migrants' wagons. Worse, three migrants drowned in the Elkhorn or nearby sloughs as they crossed.[21] As explorers, tourists, and overlanders, non-Indians interacted with the Elkhorn much as Indians had: they played the cards dealt by the river.

A new phase opened in 1856 with the founding of the town of Elkhorn, just west of Omaha near the mouth of the river for which it was named. Traveling up the Elkhorn, settlers founded the towns of West Point in 1858, Norfolk in 1870, Neligh in 1873, O'Neill in 1874, and Atkinson in 1876.[22] Town founders and settlers who followed them didn't pass through; they stuck. The Elkhorn Valley drew them up its natural highway and the river drew them to its banks. The town founders came to the Elkhorn's banks bearing technology to harness the river's energy, new ideas about how to live on the floodplain, and new ideas about how the Elkhorn ought to be.

Booster literature produced by the town founders tells us something about the environment they found (it's probably not especially reliable on this count) and even more about the town founders' dreams and ambitions, which reveals a lot about the culture and society that produced them and that they, in turn, hoped to transplant. A promotional pamphlet of fifty or so pages produced by the Elkhorn Valley Immigration Association and printed in West Point in 1880 (see Map 2-1) gives us a Rosetta Stone to understand the epic transition in human imagination that hit the basin in the late 1800s and that shapes the basin to this day.

Upon opening the pamphlet readers immediately encountered advertisements, placed by railroads, for land sales. These advertisements strongly suggest that the real estate market drove this publication. The railroads, which had received massive land grants from the US government to induce them to build, wished to sell their land. They also wanted settlers to travel on their roads and, more important, to set up homes, farms, and businesses that would create sustained freight and passenger traffic for the roads. The railroads laid their track on the natural highway along the Elkhorn and helped solve the transportation problem presented by the unnavigable river. Railroad interests believed people and products would flow easily and quickly on the river of steel that would connect the Elkhorn Basin to markets in the wider world. The Sioux City & Pacific Railroad offered discounted "land exploring tickets" to encourage potential settlers to visit the basin and shop for land.[23]

The pamphlet attacks the region's reputation as the Great American Desert and the Elkhorn is the tip of that argument's spear. Nebraska, claims the pamphlet, is a land of abundance and many river valleys, among which the Elkhorn is "*the* garden valley" (emphasis added). Not only was this *not* desert, the Elkhorn and its tributaries constituted "one of the finest and completest river and creek systems to be found anywhere between the oceans" capable of delivering "immense water power."[24] Settlers would find fertile, well-watered soil ready to produce and the hydraulic power to process the fruits of the land.

As for that fertile land, "numberless creeks" provided "perfect natural

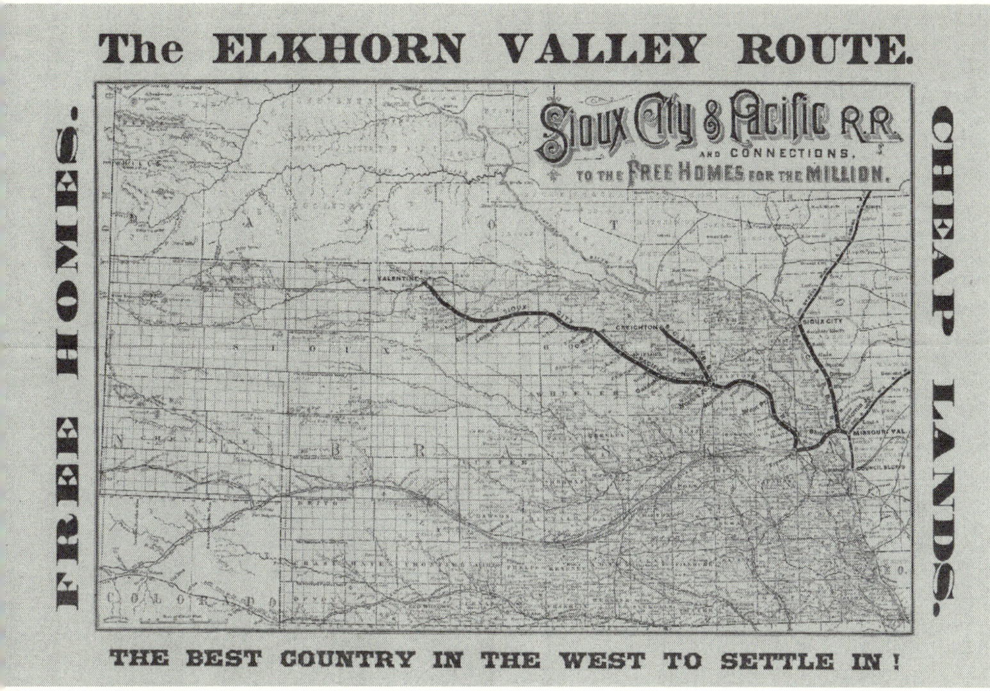

Map 2-1: Boosters, including railroad companies, tried to entice people to relocate to the Elkhorn Basin. This map from about 1883 uses a bold line running diagonally northwest to southeast to highlight a railroad route that would connect the then-remote Elkhorn Basin to the world. The railroad follows the Elkhorn's banks for nearly the river's entire length. Courtesy of Nebraska State Historical Society.

drainage." An estimated 40 percent qualified as valley land situated along streams but above the high-water mark. About 5 percent was bottomland, productive territory located between the low and high-water marks and subject to flooding. But that, the pamphlet assured, "seldom occurs, as freshets are extremely rare."[25]

Boosters promised rainfall abundant for all agricultural purposes. In the fourteen years before publication of the 1880 pamphlet, boosters claimed rainfall had increased more than 20 percent. Pioneers themselves had brought this abundance of rainfall by their civilizing actions. Planting trees, building houses, and laying railroads and telegraph lines all boosted rainfall, according to the pamphlet. But nothing boosted rainfall like agriculture. Following the best pseudoscience of the day, the pamphlet confirmed that rain followed the plow. Cultivating the soil made it more absorbent and, thus, better able to hold and then release moisture into the atmosphere by evaporation, or so the pamphlet asserted. Better still, the rainy season ran from May through July, just when crops most needed water. Furthermore, the gracious weather pattern delivered the rain in manageable thunderstorms, usually two or three times a week. About three-quarters of the storms came at night—while farmers slept—followed by clear days perfect for outdoor work. What could be better?[26]

It could be better if the landscape cooperated with the basin's beneficent hydrological cycle to deliver reliable water power, which is exactly what the pamphlet promised. For 125 or so miles upstream from its mouth, the Elkhorn's large, steady flow fell seven feet every mile. The banks rose from the water contoured ideally for town and mill sites. Even the larger creeks that fed the Elkhorn provided a steady flow of water sufficient to drive machinery and sites for mills or factories every few miles along their banks. The Elkhorn and its tributaries "formed one of the best water systems in the Union," one that could fuel agricultural production and simultaneously support economic diversification in the form of manufacturing.[27] In a flourish of booster optimism and alliteration, the pamphlet predicted that within a generation the banks of the Elkhorn and its tributaries "will be gemmed with bright, busy, commercial and manufacturing towns, in which the music of the lathe and spindle will keep time to the rhythmical flow of the waterfall and the mighty murmur and measured march of human industry and human progress."[28]

Even amidst such rosy optimism, the boosters acknowledged the Elkhorn Valley had some drawbacks. Perhaps they thought that admitting some flaws, even if they dismissed them, would lend credibility to their accounting of the valley's assets. For example, a settler might encounter "croakers and nomads," the pamphlet's term for malcontents or whiners. This population grew smaller by the day as some returned east and others drifted west. Grasshoppers once plagued the valley, even recently. Boosters dismissed stories of grasshopper damage as exaggerations and even said grasshoppers themselves were a thing of the past. Were they to return, the valley's farms would be so productive that they could feed any "'hoppers"—as well as the valley's own farmers and "all the poor of New England." Drought? A thing of the past. Hot winds? Never known to blow in the Elkhorn Valley. Floods? Extremely rare. All told, boosters promised "There is no section of the Union so free from drawbacks and so easy of settlement as the Elkhorn Valley."[29]

This pamphlet, of course, must be regarded with suspicion as far as it serves as a record of the Elkhorn Valley's climate and resources. But, it tells us a great deal about what boosters thought prospective land buyers wanted to hear. Boosters and settlers lived in a world of market capitalism dominated by agriculture in an industrializing nation. They saw, or wanted to see, in the Elkhorn Basin a promising landscape awaiting transformation. The Elkhorn had wonderful resources and a gentle, cooperative climate that was improving thanks to the civilizing activity of pioneer settlers. The benevolent, fertile Elkhorn Valley beckoned to more settlers to partake of the garden where a discouraging word was never heard. By 1880 much of the valley had railroad service. The rails promised to bring great mineral wealth through the valley as it traveled from mines in Wyoming, Montana, and Idaho to the rest of the world. The rails promised to deliver the basin's expected agricultural and livestock boom to the markets of the world and to bring the world's products to the strong-armed, clear-headed, hard-working sons and daughters of the Elkhorn.[30] The Elkhorn itself would drive that prosperity through mills on its banks. And if the Elkhorn couldn't support navigation, it provided the landscape suitable for a river of steel that would link it to the world. Unfortunately, those mills, town sites, and roads sat within the river's reach.

Floods Become Unnatural Disasters

When they placed assets in the floodplain, the town founders, entrepreneurs, and settlers set themselves up for both success and problems. With respect to the river and flooding, they made a series of choices that turned a part of the Elkhorn's customary functioning into what humans would conceive of as natural disaster. They really can't be faulted, given the technology and mindset of their day. How else was one to drive a flour mill in that time and place if not by water power? Still, the historical moment when these new arrivals planted their dreams and the machinery to make them happen in the Elkhorn's floodplain marked an important transition. Henceforth, when the Elkhorn occasionally did what it had been doing for countless generations and occupied its floodplain, that natural action would be considered a natural disaster. In truth, though, human choices produced what ought to be called unnatural disaster. Or, if we prefer to keep humans within the realm of nature, human-made disaster. The dialogue between people and the Elkhorn changed in the 1850s and that conversation remains much the same to this day.

The Elkhorn gave no honeymoon to the newest arrivals, but greeted them from the first with flooding, which turned out to be a common, if unpredictable, occurrence. In 1857, Uriah Bruner, a lawyer from Pennsylvania, helped found the Nebraska Settlement Association, which he represented that year in a journey to the Elkhorn Basin prospecting for a town site. Like the Omahas and the Pawnees who still followed the valley on their buffalo hunts, Bruner ventured up the valley struggling over flood-damaged roads to reach the town site.[31] He laid claim to a town site and began constructing a sawmill. Complications surrounding the sawmill's completion frustrated Bruner, who in 1858 sold the site to four other speculators.[32] John D. Neligh, J. C. Crawford, and two brothers named McKirahan bought the unfinished sawmill and rights to the town site.[33] Believing it to be the westernmost point of settlement at the time, the founders named the town West Point.[34]

Having overcome a flood to found the town, settlers built a sawmill and later a flour mill to put the Elkhorn's current to work. Milling played such an important role in settlement and the human dialogue with the river that it merits some discussion. Whatever the booster literature might have said about the valley's ideal mill sites, millers confronted a different reality. Nebraska's generally broad rivers made it difficult and expensive to dam them to harness their power. In addition to topographical challenges, spring flooding and sudden high rises thanks to heavy rains stood as obstacles to developing hydropower in much of Nebraska. Given their limited means, the earliest mill builders preferred rivers with well-defined channels and rock beds or tributaries that were smaller and perhaps offered better-defined channels than did main river courses. Although the Elkhorn has a sandy bed, which made it difficult to anchor dams and millworks and created an ill-defined channel that could and did move, millers built three flour mills directly on it. By contrast, entrepreneurs put eighteen mills on Elkhorn tributaries that boasted features better suited for milling.[35]

Because settlers needed to grind cereals into flour and other products, mills provided a nucleus for many settler communities and in the late 1800s in the Elkhorn Valley (and other places) that meant settlements followed rivers, the all-important power source. Farmers needed to get to the mills, so some of the earliest territorial roads were built to get them there with grain and home with flour. Mills acted as social centers where farmers exchanged news and gossip while waiting their turn. Mills also served as civic centers for polling and court sessions. They hosted religious services. Because mills were so central to people's lives, other services often located near the mills. Settlements and towns naturally grew up around mills. When Nebraska became a state in 1867, flour milling led all other industries.[36] Because mills often relied on water power, proximity to a mill meant proximity to a river. Proximity to a river meant settlements not only sat within reach of the river's milling power, considered benevolent, but also sat within reach of a river's flooding power, considered destructive, even malevolent.

While the Elkhorn and its tributaries held great promise for powering mills, rarely could millers use the water courses in their natural states; they had to be improved, which in milling terms meant they had to be dammed. A mill's power came from the fall of water and the nature of many Nebraska rivers, including the Elkhorn, required a dam to back up a head of water. The bigger the head, or vertical drop from the water level behind the dam to the water level below it, the more power could be generated as the water passed over a wheel. An 1870s-vintage, thirty-five–inch Leffel turbine, an example of the technology settlers brought to harness the Elkhorn, could generate twenty-seven horsepower with a six-foot head. Raising the head to eight feet brought more than double the horsepower out of the same turbine. Large, permanent dams cost a lot to build and many pioneer millers couldn't afford them. Instead, they turned to brush dams, which could be made quickly and inexpensively by placing logs in a stream's bed. Furthermore, sandy-bottomed rivers such as the Elkhorn did not provide a good foundation for concrete dams. Brush dams, while effective, were also easy to dislodge and a big rise in the stream or river would do just that.[37]

So it was that entrepreneurs built a brush dam to power millworks at West Point in 1859. The dam's ten-foot head could generate an estimated 500 horsepower. Some thought it the best water power in the state. Neligh and Bruner built a sawmill in 1859. Neligh added a flour mill in 1868, making it only the second mill north of Omaha at the time. Eventually drawing customers from hundreds of miles, the mill made West Point a commercial hub that provided a valuable service attractive to potential settlers.[38] Neligh and Bruner organized the West Point Manufacturing Company to diversify further. On the strength of the Elkhorn as a power source, Neligh and his partners billed the town as "the Lowell of the West," a perfect site for factories that they hoped would rival the Massachusetts factory town whose name they invoked.[39] For a while, it seemed the vision might come to pass. At the company's peak, the Elkhorn powered a planing mill, a paper mill, a furniture factory, and a condensed milk operation with a ten-thousand–can per day capacity, and a brickworks.

At one point the company's flume guided the Elkhorn's working waters over thirteen turbine wheels. In the late 1880s, when West Point built its first waterworks, hydro-power supplied by the mill complex pumped water through the system's five miles of mains to homes, businesses, and fire hydrants in town.[40] But as the Elkhorn worked milling equipment, so it worked on the dam, which required regular maintenance to the tune of about $1,000 a year. Financial problems plagued the company and one by one the turbines came to a halt until only the flour mill continued to turn in 1900.[41]

As the quick establishment of the sawmill indicates, early settlers placed a high value on timber as a construction resource. An early history of nearby Antelope County describes the countryside as "handsome, inviting, rolling, prairie land, well watered by numerous streams, and a belt of timber skirting the Elkhorn River and most of the creeks, with occasional groves of considerable size."[42] Within a year of Bruner's arrival and West Point's founding, its sawmill began devouring the native timber, mostly cottonwood, white elm, ash, and several kinds of willow, converting it into planks and siding for settlers' houses. Aside from the trees growing along the Elkhorn and its tributaries, however, the region was devoid of timber.[43] By 1872 West Point's voracious appetite for lumber nearly exhausted the local supply of trees.[44]

In Madison County, early settlers sought to relieve the prairie's monotony and to provide for their future needs by planting hundreds of groves of trees. In addition to shade, lumber, windbreaks, and firewood, these settlers apparently believed in a corollary to the once popular maxim "rain follows the plough." In this case, they hoped rain would follow the bough.[45]

A story from the logging days illustrates how the Elkhorn's nature dashed settler dreams of remaking the basin into a profitable, wooded landscape. Nine local entrepreneurs set out to harvest timber on government land, such timber being considered common property. They made rafts out of the logs and floated them down the Elkhorn to a sawmill. The men abandoned their work on about July 1, 1871, to replenish their supplies and attend a nearby Fourth of July celebration. When they returned on July 5, the river had dropped eighteen inches, which made rafting impossible. Floating the logs singly, they were able to get them to the mill, but not without frequent hang-ups on sand bars. Upon reaching the sawmill, however, the logs' bark was so impacted with sand that the miller refused to saw them until the bark was removed. When the loggers exhausted the local timber several years later, mill owners sold their machinery, which was moved out of the county.[46] The newcomers' imaginations

Figure 2-1: This scene on Logan Creek, pictured on a postcard mailed in 1913, shows how settlers manipulated Great Plains waterways to generate power. They also enhanced recreational opportunities as evidenced by the person fishing on the right. Author's collection.

and technology, powered by the Elkhorn, stripped the countryside of a rare commodity and exposed the limits of one sector of an extractive economy. The Elkhorn, though, proved an unreliable mode of transportation, even for floating wood, and one with a very gritty drawback.

Some mill owners eschewed diversified manufacturing, but capitalized on the foot traffic generated by their mills to operate other businesses. At Norfolk, Charles Mathewson, who built his mill in 1869, opened a general store at the mill. During the busy immigration season of 1870 the store registered daily sales of about $100. The owner of Logan Mill added a blacksmith shop near his Logan Creek site.[47] A photograph of the mill dam on Logan Creek at Oakland, Nebraska (see Figure 2-1) shows not only how the dam enhanced the creek's potential energy for the purpose of milling, but also suggests how a mill dam created a mill pond and a fishing hole.[48] The mills, then, provided entertainment and recreation as well as industry in the river's floodplain.

In 1872, Neligh bought land farther up the Elkhorn to found the town in Antelope County that bears his name. The following year he established a water-driven flour mill and had a frame dam built across the Elkhorn to divert water through the mill's sluice. He sold the mill in the financial panic of 1873.[49] By 1872, settlers in Antelope County employed two sawmills, one

Figure 2-2: This view of the Neligh Mill from about 1915 shows the advantages of its location and the symbiotic relationship between the Elkhorn (out of view just beyond the mill), resource processing initially powered by the river, and transportation routes. Courtesy of Nebraska State Historical Society.

steam-driven and one water-driven on Clearwater Creek, to process timber harvested locally.

Water-powered technology allowed people to harness part of the Elkhorn's flow to process raw materials and to bring manufacturing to the basin's economy. They turned water power into imagined horsepower to transform the valley's cereal crops into food for people and feed for livestock. Those calories turned into human and animal work in the valley and beyond. The Elkhorn's power also drove a basic industry in the valley and allowed the residents to participate in national and international markets as railroads carried flour and feed out of the valley along the same route followed by and shaped by the Elkhorn (see Figures 2-2 and 2-3). That fraction of the Elkhorn's usual flow that created horsepower to turn heavy milling equipment also offers a glimpse into the exponentially greater power that a bank full or overflowing Elkhorn brought to bear on whatever happened to be in its floodplain at the wrong time. That horsepower could mill and it could just as easily rend and kill.

Figure 2-3: The Elkhorn provided the power to mill these sacks of Gold Medal flour stacked in the Neligh Mill warehouse in 1911. Flour left Neligh by rail and went as far as Europe. Courtesy of Nebraska State Historical Society.

The spring of 1873 brought possibly the first truly and certainly the best recorded early catastrophic flooding of the Elkhorn's early settler colonial period. Newspapers had come to the basin and their reporting left the richest record of Elkhorn flooding to that time, revealing a floodscape much changed in the course of a single generation. Snow and ice transformed the Elkhorn into a menace. An April blizzard dumped enough snow to smother livestock. Visiting Pawnees ate some of the animals, which helped dispose of carcasses. The storm also dumped enough snow to dam portions of the Elkhorn, which made local news. The *West Point Republican* reported that the snow dams caused extremely low flows in areas downstream from the obstructions. The pent-up Elkhorn also created a disaster waiting to happen. "Look out for a rush of water soon," the *Republican* warned readers.[50]

Continued warnings of impending floods and comments about high water in Neligh and Norfolk appeared throughout May. By May 29, the Elkhorn flowed at its highest level in six years. West Point residents "risked their lives . . . working like beavers" to clear driftwood from the bridge across the swollen Elkhorn.[51] The bridge, normally a boon to the local residents, became a driftwood trap, which effectively turned it into a dam. The water held back by the bridge-dam had nowhere to go but into town, parts of which were under water for the first week of June. On Tuesday, June 3, the Elkhorn ran over the east side of the West Point mill dam and continued to rise until it reached the railroad tracks. By Wednesday, June 4, several businesses and residences were flooded and two or three families abandoned their homes in boats. Other residents were able to pilot boats over the stretch of prairie, now under water, that had separated the town from the river. On June 5, the *Republican* reported "much damage" to crops, but little damage other than to gardens in town.[52]

Receding waters revealed a different story, one of extensive devastation. Mill dams suffered significant damage, as indicated by a report in the *West Point Republican* that West Point's mill dam and bridge withstood the high water pressure better than any other in the area.[53] The record of damages suggests how dramatically the Elkhorn's floodplain had changed thanks to new settlers implementing their dreams in the basin. Dams had been placed in the floodplain and had been damaged. A bridge had been built over the river to make travel easier, but it trapped floating debris, which turned it into what was probably the biggest dam on the river at West Point. It very effectively, if unintentionally, pushed damaging floodwaters into West Point, the construction of which, along with related structures, had redefined the contours of the floodplain. So, even if the town had been placed above what the booster literature classified as bottomland (or outside the pre-settlement floodplain, to put it another way), development changed the river's boundaries. Houses and businesses sat in the floodplain, as did the recently arrived river of steel, which found itself threatened by the river of water (see Figure 2-4).

Downstream the Elkhorn did more than threaten—it damaged a railroad bridge on the transcontinental Union Pacific line that led to a train derailment in what might be the most bizarre episode in the river's flood history. Well aware of recent heavy rains that had swollen the Elkhorn, the engineer of the west-bound passenger express eased the train onto the 400-foot bridge at two to three miles per hour. At about one hundred feet onto the bridge, the trestle's timbers collapsed. The engine, tender, and first car dropped about ten feet into a slough with about ten feet of water flowing rapidly into the Elkhorn. The engineer escaped the cab by swimming through an open window, but another railroad employee, pinned in the cab, drowned. Insofar as those details go, the accident qualified as tragic, but otherwise unremarkable. The third car that fell into the Elkhorn elevates the accident to the bizarre.

That third car, the "celebrated California Aquarium car" as the news reports called it, plunged into the water carrying Livingstone Stone, US fish

Figure 2-4: Railroads such as this one running through West Point and shown on a postcard mailed in 1909 connected the Elkhorn Basin to the world. The roads followed the river, but the river's floods occasionally stopped traffic. Roadbeds and bridges altered the Elkhorn's bed and sometimes exacerbated flooding. Author's collection.

commissioner, two assistants, and a dozen or more tanks loaded with fish, lobsters, and oysters bound for California. Californians had requested the fish to stock western waters and, perhaps, for sentimental reasons. A report of the crash speculated that its news would disappoint Californians who longed to see again fish they knew in their childhoods back east. When the car hit the slough, its cargo, estimated to be worth $30,000, escaped the tanks and found a new accidental home. This came as good news to Omahans, who considered the Elkhorn cheaply stocked with fine fish. The fish likely thought differently. Black bass and glass-eyed pike had come from Lake Champlain. Bull-heads and yellow perch had come from the Missisquoi River in Vermont. Silver eels and shad eggs had come from the Hudson River. Salt-water eels, tautogs, and lobsters had come from the Atlantic Ocean near Martha's Vineyard, and oysters had come from Massachusetts Bay. The tautogs, oysters, and lobsters faced quick death in the Elkhorn's fresh water. Some of the other species, though, might have thrived as accidental settler colonialists in a Plains river.[54] Meanwhile, human settler colonialists in and around West Point busied themselves repairing roads, worrying about disease spreading from large quantities of stagnant water in town, and probably rejoicing that crop damages turned out to be “hardly worth mentioning.”[55]

The 1873 flood followed a pattern that would become familiar to subsequent generations of Elkhorn residents. Snow and ice dammed the river, which could cause flooding behind the dam as well as create the potential for a damaging surge downstream when the dam broke. The flood also called attention to changes that had come to the basin with the arrival of a new culture and society. The river’s ability to drive mills drew settlers to its banks, where the river’s flooding punished them. The valley’s level topography made it an attractive site for roads. But roads couldn’t only run along the river, they had to cross it from time to time, which required bridges. The roads and bridges benefitted the settlers, but placing those assets in the floodplain made them vulnerable to destruction by flooding. Furthermore, a bridge, if it collected flotsam and jetsam carried by the flooding river, could turn into an unintentional dam with damaging consequences. The freakish wreck of the aquarium car can be seen as another illustration of the American imagination at work. The same collective mind that brought milling and damming to the Elkhorn, also introduced exotic aquatic species to the river. People had not intended those aquatics for the Elkhorn, but they had intended them, apparently, as either a comfort for eastern migrants to California or as an “improvement” for western waters or both. People in Omaha saw the fish as an improvement to the Elkhorn, even if a reporter had written about it tongue-in-cheek.

The rest of the 1870s passed without major flooding, but not without incident. High water in the spring of 1875 reached within six inches of the floor of a railroad bridge (likely the same bridge that washed out in 1873, although the report isn’t clear) and high water and ice swept away at least four county bridges. In the spring of 1877 both the Platte and the Elkhorn rose. The Platte threw up a bank that effectively dammed the Elkhorn, causing the latter to rise even higher near its mouth. The Elkhorn jeopardized some five thousand acres of cropland, washed out a dam at West Point, and claimed the lives of two women who drowned after their boat capsized. A photographer recorded the 1877 Norfolk flood, with water covering Norfolk Avenue and lapping at the street-level entry of a hotel and a hardware store (see Figure 2-5). And in 1877 a report surfaced that anglers could catch eight-pound salmon in the Elkhorn, where they flourished. The salmon, not native to the river, allegedly came from the aquarium car that had fallen into the river in 1873, although contemporary reports did not list salmon among the species riding in the car.[56]

Figure 2-5: The flood of 1877 covered Norfolk Avenue in Norfolk and left people and horses to seek high ground near buildings. Courtesy of Elkhorn Valley Museum.

The Elkhorn's notable floods of the 1880s and 1890s followed familiar seasonal rhythms and continued their contentious dialogue with the basin's railroads. In late March 1881, the Elkhorn climbed more than six feet above its low water mark at West Point and stood poised to climb further. Ice blockages seemed the most likely cause.[57] At Waterloo, near the Elkhorn's mouth, a four-foot flood surge lifted the Union Pacific railroad bridge there several inches off its foundation. Surging water washed out Hooper's wagon bridge snapping the two-inch cable rope supports like whip cords. River watchers predicted the Elkhorn might take two or three weeks to return to normal.[58]

The Elkhorn proved them right and continued to inflict damage through much of April. The April 21 *Fremont Herald* reported that the Elkhorn "dashes through the streets of Norfolk at will" doing much damage, but making for wonderful boating through town. In the flooded area, Norfolk's streets lay submerged under two to three feet of water and the river washed out more than a half mile of railroad track upstream from the city. Two railroad bridges in the same vicinity also fell to the flood, cutting train service to the end of the line. The flood damaged approaches to the railroad bridge at Neligh, but the bridge itself held. The Elkhorn washed out the Union Pacific crossing at Waterloo in late April halting transcontinental traffic for several days. As of April 25, Waterloo sat under five feet of Elkhorn water as did farms for several miles along the river. Area residents blamed the railroad embankment for "throwing water back" on them. They threatened to breach the embankment to create a drainage channel. Before they could act, a sheriff and posse of constables arrived from Omaha to protect the railroad.[59]

These floods show that early white settlers adapted to flood conditions while waiting for the waters to recede and then rebuilt their damaged structures—railroad lines for instance—in the same place that had just succumbed to the river.[60] Hooper's innovative residents restored wagon travel by pooling their money to buy lumber to build a boat. They twisted fence wire into a cable and stretched it across the river, and began ferrying wagons and passengers across the swollen Elkhorn.[61] The floods also offer insight regarding tension between "little people," the inundated denizens of Waterloo and surrounding farms, who perceived themselves aggrieved by floodplain modifications constructed by the Union Pacific and protected by law enforcement officials. The Elkhorn's floodplain proved again to be a risky, contentious place and the site of disasters that were anything but natural.

The spring of 1888 saw notable flooding hit the Elkhorn Basin again, this time in a rare upper-basin flood that affected areas as far upstream as Holt County. On May 26 and 27, a Saturday and Sunday, runoff from heavy rains brought the upper basin's streams over their banks. Bridges between Clearwater and Neligh washed out halting rail traffic westward out of Neligh. The Elkhorn continued to rise until that Tuesday night when floodwaters reached their highest point since 1881. At Neligh, only a ripple in the swollen river's current provided any sign of where the mill dam was.[62] High water and ruined bridges slowed commerce in the town for days.[63]

As of Monday the 28th, Norfolk remained dry as levees along the North Fork contained the stream, which had climbed to within eight inches of the levee's crown. Like the Elkhorn at Neligh, the North Fork climbed to its highest level since 1881. Between 3 a.m. and 4 a.m. on Tuesday the 29th, the North Fork tore a seventy-foot crevasse in the levee and swept into Norfolk as far west as Main Street. Water ran deep enough for boating in the neighborhood's gutters. Residents scrambled to plug the gap with 600 sandbags and later that day the flood began to recede. Streets and sidewalks suffered minimal damage, but houses flooded, a length of railroad track washed out, the county bridge south of town was impassable, and water covered several farms south of the Elkhorn.[64]

In response to the flood, a group of ingenious and civic-minded Neligh residents collected money to build a four-oared flatboat to ferry passengers across the Elkhorn at no charge. The town council also authorized a bank-stabilization project at the town's waterworks pumping station, built near the Elkhorn, to prevent future floods from washing the building downstream.[65] Norfolk's *Journal* credited the levee on the North Fork, built just upstream from the mill dam, for minimizing the effects of the 1888 flood. But, it also claimed that a larger, better-built levee would have kept the "raging Elkhorn within its proper bounds." The *Journal* renewed its call a week later urging the town to enlarge and strengthen the levee to prevent future flood damage.[66]

The 1888 flood tells us much about the relationship between the pioneer Anglo settlers and the Elkhorn. Newspaper reporting referred to levees. Clearly by this point people at Norfolk recognized that flooding occurred regularly enough to warrant the expense and trouble of building defensive earthworks to protect assets they had placed in the floodplain. Maybe some of them even questioned the decision to place a town and its significant assets in the floodplain. Having made that decision, though, Norfolk residents doubled down by going to the trouble of building the levees. Building a levee required a significant amount of work, which indicated that Norfolk had reached a point where it had the population, wealth, and technological ability to undertake the task. The act of constructing the levee further revealed a worldview confident in its ability to remake the landscape and to control the force of nature that was the Elkhorn. As the newspaper report said, the people of Norfolk had formulated a notion of where the Elkhorn should flow and they were determined to keep it "within its proper bounds."

Downstream from Norfolk, a pair of summer floods in 1891 exposed the limits of hydraulic engineering projects at West Point. At about 3 p.m. on June 24, lightning ripped through the summer sky over Cuming County. Thunder rumbled over the horizon as sheets of water fell for five hours from a thunderstorm that dumped seven and one-third inches of water on West Point and surrounding areas. Storm runoff transformed West Point's east–west streets into streams flowing through town. Buggies, wagons, cultivators, mowers, lumber, and other flotsam and jetsam floated past downtown buildings toward the Elkhorn. The waters washed away the newly built gutters meant to channel them and unceremoniously dumped the gutter bricks in fields outside of town. The flood's natural waterworks also laid bare the pipes of West Point's man-made waterworks. Only a few Main Street businesses escaped damage.

Alarmed at the plight of citizens who lived near the river, a rescue squad set out in wagons and boats to bring the lowlanders to safety. Refugees stayed overnight in the city hall, the courthouse, hotels, and in private homes. Damage estimates put the city's loss at $10,000, private losses in town at another $10,000 to $15,000, and county losses at an additional $40,000. Less than one month later, on July 23, another flood, reportedly worse than the first, "filled the basements and did much damage."[67] West Point's residents evidently had not gone to the lengths of Norfolk's in protecting themselves from the Elkhorn. Yet clearly they had given thought to managing water.

Ironically, perhaps, the 1890s saw the importance of water-powered milling, the very thing that had drawn newcomers into the floodplain, hit its peak and begin to decline. Expanding railroad networks allowed millers to import steam engines that could free them from the Great Plains' fickle and dangerous rivers. Nebraska's first millers *had* to build on sites with favorable water power. But favorable is a relative term. Droughts could and did reduce stream flows and, in turn, horsepower. Rivers could and did change channels, sometimes robbing a water-powered mill of its hydro-horsepower. Worst of all, the spring rise could and did damage dams, turbines, and even the mills. The rise might not destroy equipment every year, but when it did, repairs and lost capacity cost millers dearly.

Railroads killed some water-powered mills because of their location. The Neligh mill was lucky in the sense that the railroad followed the main line of the Elkhorn, which brought rail service to the mill's door and extended its working life. Wells Mill on Shell Creek didn't fare as well when the railroad came through Schuyler, two and one-half miles from the mill's location. Hauling wheat and flour over that distance didn't make sense, so the mill closed and moved to town where it would be close to the tracks.

Changes in the land brought other threats to water-powered mills. Early settler accounts from eastern Nebraska suggest that waterways enjoyed stronger and more reliable flows before the sod was broken for crops. Soil from plowed fields washed into creeks and rivers. The silt load made some waterways unreliable for milling and forced the abandonment of some mills. The silt also raised riverbeds and made flooding more destructive.[68] The mills, then, that had brought the newcomers and their assets into the floodplain were becoming increasingly irrelevant. But the towns and the roads they had inspired remained in the floodplains.

Conclusion

As in many other river valleys, the Elkhorn and its immediate vicinity—the floodplain—invite human use. But there has always been a dialogue of sorts between the river environment and the human cultures using it. It could be said that the earliest settlers, from the Paleo-Indians to the Omahas, listened to the river. They built their homes above the floodplain and likewise kept most of their possessions out of the river's reach. But as non-Indian settlers arrived with new technology and new ideas about how to apply it in their colonization of the Great Plains, the dialogue between settlers and their environment has changed with humans doing more of the "talking" and increasingly less "listening."

In the 1800s the countryside tacitly expressed the potential of easy overland transportation, hydraulic power for mills, fertile farmland, and accessible groundwater. Humans responded by building wagon roads and railroads. They harnessed the river's power to run mills, pump water, and generate electricity. Commerce evolved luring more settlers, spawning towns, and leading to further intensification in the use of local resources. Waterworks brought rivers into homes and factories and allowed humans to start and stop the flow of "indoor rivers" at will. In time the Elkhorn's mercurial ways, the unreliable nature of the river's water power, and the advent of other more reliable energy sources diminished the river's importance to settlers who paid less attention to it. But the river's power had opened the door to new cultural patterns. As non-Indian settlers moved into the region, many of its former residents moved out. The Omahas and members of other tribes left or were forced out. Game animals became rare sights. Herds of bison once trampled the valley's sod, but Stanton County's early US settlers recorded sighting only carcasses. In Antelope County, Stanton's neighbor to the west, reports indicate the last buffalo was killed in 1872 by an Indian hunting party. Elk, deer, and antelope reportedly roamed Stanton County into the 1870s. By 1892, however, a historian tellingly, even if in exaggeration, wrote, "I doubt if a dozen wild deer could be found in Stanton County."[69] The basin's scant timber briefly supplied a local lumber industry, but that boom busted quickly when the trees ran out and the sawmills closed. Intensified agriculture led farmers to supplant native grasses with row crops that disturbed the environment's natural hydraulic cycle and its built-in checks on runoff. Bridges and roads that carried people and goods in and out of the basin also altered the Elkhorn's floodplain and the river's flood mechanics. Intentionally, accidentally, and with unintended consequences the Elkhorn Basin had by the end of the nineteenth century become a place where the river's natural actions had become unnatural disasters.

Chapter 3

IMPERFECT PROGRESS AND THE ELKHORN: FLOODING AND FLOOD FIGHTING, 1890S–1930S

"Expected Flood Arrives"

Headline in the *West Point (NE) Republican and Cuming County Advertiser*, April 5, 1912

By the 1890s, a single generation of newcomers had dramatically changed the Elkhorn River Basin. New settlements dotted the basin, many of them organized around mills that clung to the banks of the power-giving river. Railroads connected the basin to the rest of the world with unprecedented ease and speed. As the basin's new inhabitants would have seen it, they had overcome a great deal in their quest to improve a "blank" landscape that now accommodated industry and large-scale agriculture. In short, they had brought civilization and progress to the basin. Still, they were married to the river in many respects. The railroad afforded them the beginnings of independence, but many mills continued to rely on water power to keep their gears turning and the settlements and railroads built around those mills sat on ground in the floodplain that, according to one way of thinking, truly belonged to the river. But that way of thinking did not prevail at the time. Rather, the dominant attitude promoted perfection of the environment to suit civilization and the notion that "problems" inherent in wilderness or untamed landscapes could be corrected.

Descendants of the newcomers continued on their course of improving the landscape in ways that resonated with their conception of progress through the first decades of the twentieth century. Historians have labeled this period the Progressive Era and its themes played out on the national stage in various ways. With respect to the environment, policy often aimed at taming and rationalizing landscapes. In the American West, the Progressive Era witnessed the birth of the big dam era, which ran through the first half of the twentieth century. Engineers used dams to regulate rivers, to control drastic swings in flow as seen in flooding, to impound water for irrigation, and to generate electricity. The era also saw new approaches and newly aggressive measures to manage rivers for navigation. These approaches found their way to rivers on the Great Plains, although not all were suitable for large-scale modification projects. The Elkhorn, twice removed from the mighty Missouri (which was modified for navigation, flood control, irrigation, and hydroelectric generation), fell into that category. Progress and the Progressive Era came to the Elkhorn Basin on a smaller scale. Millers began converting to alternate forms of power that didn't depend on the unpredictable Elkhorn. Residents undertook new and more aggressive measures to protect their land and other property from the Elkhorn's occasional high water. From the 1890s through the 1930s, valley residents sought to separate themselves from the river but found they could neither sever strong ties established by the pioneers nor conquer their watery neighbor.

Water and the Progressive Era at Neligh Mill

The biography of the Neligh Mill reflects several Progressive Era themes and also shows how the human–river relationship remained close and intimate during the first decades of this period, but began to fray. Two photographs of the Elkhorn preserved on period postcards vividly illustrate the intimate connection between the people and the river that continued into the early twentieth century. One image, made in about 1907, shows a riverbank landscape developed to work (see Figure 3-1). The head gate to the mill race, or the channel that takes impounded water to the turbines, occupies the lower left foreground. That is where humans have manipulated the river to harness its energy. A bridge spanning the Elkhorn sits center left in the image. The bridge allowed ground transportation (animal-drawn wagons and later trucks and automobiles) to overcome the obstacle of the Elkhorn. This, too, represents an intimate connection between the people and the river. Both the bridge supports and the mill race literally touch the Elkhorn and connect humans to the river's floodplain or directly to the river itself. The image's background shows how little timber grows on the Plains in areas near rivers and streams (left background) and how humans have cleared some native timber to plant crops

Figure 3-1: This view of the Elkhorn from about 1907 photographed looking down from the Neligh Mill shows the river both as a transportation obstacle to be bridged and as a source of energy to power the mill. Portions of the mill dam can be seen in the foreground, as can the head gate to the mill race. Floods occasionally washed out the dam and would eventually destroy the head gate, mill race, and bridge. Author's collection.

and buildings on or near the floodplain (right background). The image shows a landscape that has been improved according to the agenda of US society and its market economy. It is a picture of progress.[1]

A second image, a photograph of Neligh Mill made by Neligh photographer Emanuel Wolfe probably before 1911, shows the millworks perched on relatively high land in the background (see Figure 3-2). In the foreground and in the river sit the concrete dam, critical to backing up the Elkhorn, and the mill race and power house that allowed the Elkhorn to drive the mill's turbines. In addition to what the images tell us about the Elkhorn's development, they also suggest the importance placed on the river by those who lived there. Photographers took pains to record these images and the Wolfe photograph appeared on a postcard published by Payne & Shanes Pharmacy, a Neligh business that presumably wished to promote a notable local attraction.[2] This, too, is a picture of progress and an example of how Elkhorn Basin settlers improved, or brought progress, to what would have been considered a wilderness.

Details about the mill's working life at the dawn of the twentieth century flesh out our understanding of those images. The mill processed wheat, corn,

Figure 3-2: Probably made before 1911, this photographic postcard shows the intimate bond between the Neligh Mill and the Elkhorn.
Courtesy of Nebraska State Historical Society.

and oats for human consumption and for livestock feed. By 1899 Neligh boasted one of Nebraska's top grain-handling facilities. Operators added warehouses in 1886, 1900, and 1909, suggesting that the mill was growing as more people moved to the basin and as the railroad expanded access to raw materials (the mill purchased wheat from farmers in Nebraska and South Dakota) and to markets. While the railroad helped Neligh Mill, at least at the time, it killed some others. As mentioned in Chapter 2, Wells Mill on Shell Creek closed and moved into Schuyler where it could be close to the tracks. Things were different at Neligh. In its earliest years, the mill supplied markets in Nebraska and South Dakota. In the 1890s Neligh Mill won US government contracts to supply Indian reservations and military installations. After the turn of the century, the mill's products traveled as far as England and Scotland.[3]

Stephen F. Gilman, who bought the mill in 1898, oversaw much of the mill's dramatic expansion and diversification, which he accomplished, at least initially, with hydropower from the Elkhorn. Gilman replaced the brush dam with a larger one of concrete, which created a nine-foot head and allowed the mill turbine to generate more power. The new, higher dam expanded the mill

Figure 3-3: A group portrait of Neligh Mill workers made in 1911 shows men whose livelihoods derived in large measure from the Elkhorn's power.
Courtesy of Nebraska State Historical Society.

pond, which required construction of dikes and a bypass spillway to manage the Elkhorn's waters. Gilman applied some of that newfound power to generate electricity, which he sold to the city of Neligh. He capitalized on the existence of electricity to further diversify the mill's operations by selling electric appliances. Mill employees wired homes and businesses for electricity and strung electric wires from the mill to customers in the town. In doing so they literally brought electricity to Neligh and that modern power came from the Elkhorn. In these years the Elkhorn perhaps had its deepest reach into the lives of people living in the basin. It not only ground wheat into flour that they used in their kitchens, but also provided the electricity that powered new labor-saving technologies. Edison Gem Lamps (light bulbs) sold at the mill and powered by its electricity lit homes. Triangle Lektrik Irons, also sold at the mill, promised to make ironing better, cheaper, and easier. Perhaps most revolutionary, if its advertising copy is to be believed, the Frantz Premier Electric Cleaner (vacuum cleaner) turned housework into play. "The hard work of sweeping and dusting is no longer to be praised—it is misdirected energy and wasted time," promised the Frantz Company. The cleaners could be purchased, of course, at

Figure 3-4: The flood of 1897 covered the streets of Norfolk. Based on figures standing in the water, the flood at the time of the photograph was about ankle-deep. But, the bottom row of windows on the *Norfolk Daily News* building suggests how water ankle-deep in the street could be four feet deep in the building's basement. Courtesy of Elkhorn Valley Museum.

the Neligh Mill Light Department. The Elkhorn, water flowing downhill, provided jobs, processed the basin's agricultural bounty, and powered the modern labor-saving devices that promised a quality of life better than that enjoyed by the pioneer generation (see Figure 3-3).

The Elkhorn generated electricity at Neligh until 1915, when Gilman turned to a diesel generator. This signaled the beginning of independence from the Elkhorn's motive power, but the hydropower era at Neligh would last until 1920, at least when it came to milling flour. The story behind the abandonment of hydropower appears below. For now, it's enough to note that the new, expanded works brought changes to the mill and to life in Neligh, but a considerable portion of those works sat in the floodplain and involved managing the Elkhorn in increasingly aggressive ways. Though the mill itself, which sat on the relatively high north bank, rarely, if ever, suffered flood damage, the same could not be said of the dam and the machinery that converted the Elkhorn's energy into milling and electrical energy.[4] As this story suggests, the people of the Elkhorn Basin relied heavily on the river for the energy to modernize and to industrialize; but, as the Neligh Mill's break from water power in 1920 foreshadows, the relationship between people and the river was moving into new territory.

The Flood of 1897

As the human–river relationship evolved, flooding remained a constant in the dialogue and the first significant example in this era came in 1897. The melt of a winter's heavy snowfall brought flooding to Norfolk—twice in short succession—during the early spring of 1897 that revealed the scattered and individualistic approach to flood control that still dominated the basin as well as tensions among economic interests in the town. Norfolk's residents had a good idea that trouble stalked them as they watched the North Fork rise about three inches an hour all day and into the night of Friday, March 19. Sentinels stood watch on a railroad embankment north of town as Norfolk went to bed. The embankment and a city dike stood between the swollen river and the city's businesses and residences. Overnight the river began rising more rapidly. A long steam whistle and a fire bell woke Norfolk at 4:30 a.m. with the alarm that the breach had come. Water poured through and spread out over a mile and a half.

The damage was terrific. Much livestock, mostly cattle and horses, had been taken to high ground, but many hogs drowned. Norfolk's business district became a shallow lake. In the flooded areas water stood three feet deep, on average, and in some places as deep as six feet. Virtually every building with a basement, which was most of them, took on one to three feet of water. The Norfolk *Daily News* building had nearly four feet of water in its basement (see Figure 3-4). The water destroyed goods stored in basements with one business owner estimating that he lost more than $1,000 worth of merchandise and another more than $400 (in 1897 dollars). In many other cases, basement flooding extinguished the coal-fired furnaces that heated the establishments leaving building occupants shivering in the chilly March weather. Near the mill, floodwaters destroyed more than 8,000 bushels of corn waiting to be processed and at least 1,600 bushels of oats.

The flood disrupted travel and commerce. The tracks of the Chicago, St. Paul, Minneapolis, and Omaha Railroad (M&O, for short) came into town from the north, where they crossed the river. The washout of that line's roadbed forced incoming passengers and freight to be unloaded north of town. From there, passengers walked and freight was transported along the city's dike into town. The flood washed great ditches under the tracks in the railyard shared by the M&O and the Union Pacific. As the tracks settled, freight cars toppled and workers began repairing the road as soon as the water subsided. The flood also washed out almost every sidewalk extending three blocks from the river downtown with repairs estimated to cost hundreds of dollars. A small bridge carried downstream by the flood struck the First Street bridge over the North Fork and threatened to take it out of commission, but residents quickly disassembled the renegade bridge before it could damage the bridge against which it had wedged. Business came to a near-standstill as residents hustled to move their property to high ground, went for boat rides, or simply dropped what they were doing to gawk at the watery world their city had become.[5]

Norfolk hadn't fully recovered from the March flood when another hit in early April. The April flood proved more or less a replay of March. High water flooded almost every business in downtown Norfolk and spread out over low land east of the North Fork for a half mile. Basements flooded, sidewalks washed out, roads suffered erosion, and the high water reached within inches of marks notched just weeks earlier. Coverage in the *Norfolk Daily News* included remarks critical of city leaders that reveal telling clues about Norfolk's relationship with the North Fork. Upon hearing predictions of the second flood, the mill company put forty men to work erecting a dike to protect the property. The dike reportedly cost about sixty dollars and effectively turned water away from the mill. Had the city taken similar quick action to repair breaches in the railroad embankment (the ones caused by the March flood), the business district could have been spared. Evidently, the thought of spending public money on what might have been construed as a benefit to the railroad company kept the city from acting. That raises the question of why the private sector company didn't hasten to repair its infrastructure. It also suggests what might have been lingering echoes of anti-railroad sentiment that rang across substantial portions of the Plains during the heyday of the Populist Party. Or the newspaper might have just expressed what must have seemed economic common sense as it suggested that repairing the flood damage would cost the city ten times the bill for the preventive measures it suggested. The reporting also pointed out the state of the floodplain landscape. The railroad embankment served as part of the city's de facto flood protection. The city had erected a dike as well. And, the mill company had thrown up its own ad hoc protective structure when threatened. Norfolk and corporations residing in the city had an intimate and somewhat coordinated relationship to the river. By all accounts the floods of 1897 showed that the city's dike held. When the railroad embankment washed out, the city—and the railroad—suffered. The mill acted decisively to protect itself by throwing up a dike, which succeeded in protecting the mill, but in all likelihood did so at the expense of its neighbors in downtown Norfolk as water shunted away from the mill probably found its way onto Norfolk Avenue and surrounding properties.[6] Living with the Elkhorn's North Fork was an unavoidably intimate and soggy affair and one marked by a hodge-podge of loosely coordinated, at best, approaches in the late nineteenth century.

Progressive Organization: Drainage Districts

In the late 1800s and early 1900s, most Americans viewed flood control as a matter of self-protection. Individuals had the responsibility to protect themselves or they might band together locally. Citizens formed local levee districts and built modest, usually earthen structures to contain floodwaters. When those levees held up against customary seasonal flooding, it often inspired the levee district to expand its efforts—to build larger structures to protect more of the floodplain for development, be it agricultural, urban, or something else. The USACE steered clear of flood control, largely because its main mission was to promote navigation. Federally elected officials not only shared the prevailing sentiment that flood control was a matter for local or state government, they also had little appetite for the political and economic burdens that would accompany the task of flood control. An exceptionally destructive flood on the Missouri River in 1903 inspired a number of people who lived in or farmed the Missouri's (and some tributaries') floodplains to change their views. They began to push for federal flood protection. Local politicians and newspapers joined the cause. Another damaging flood in 1908 lent urgency to the matter. Congress responded with money for channelizing the Missouri from Kansas City to St. Louis for navigational purposes. Although Congress stuck to its mission of promoting navigation, this satisfied the flood control lobby because they believed the deeper navigational channel and related projects would also contribute to flood control. But things were changing, at least downstream from the Elkhorn.[7]

In the first decade of the twentieth century, the desire for flood control led people in several Elkhorn Basin counties to form drainage districts—private, quasi-governmental organizations dedicated to improving drainage within the jurisdiction of each. These initiatives can be viewed as a manifestation of the Progressive Era ethos running through the United States at the time. Progressivism, a rather unwieldy umbrella concept, included a range of people and organizations who identified various problems and set out to solve them using the latest scientific knowledge and, often, government support to improve the human condition. Well-known examples of progressives include Jane Addams, who established the Hull House settlement in Chicago to help immigrants and others cope with urban life, and Theodore Roosevelt, who as president supported pure food and drug measures and conservation. The creation of the US Bureau of Reclamation in 1902 applied progressive-era principles to water management in the western United States, notably in the construction of dams for hydroelectric power generation and irrigation that promoted economic development and quite literally remade the arid West. Irrigation districts along the Elkhorn reflected a similar desire to remake and improve the hydroscape, but on a smaller scale.

One anti-flood strategy was to "improve" the Elkhorn and its tributaries. So-called improvements consisted primarily of channel-straightening projects without appreciable channel enlargement. The US government began applying this strategy to the Mississippi decades earlier to improve navigation, mostly, but also in an attempt to minimize flooding. Mark Twain, maybe the most famous of the Mississippi's river pilots, called it the crookedest river imaginable. In one stretch the river took 1,300 miles to cover what a crow could fly in 675, he noted in the 1850s. In a few decades of work by the USACE, channelizing and straightening, the river lost hundreds of miles and with them the character of particular bends, wetlands, and plant and animal communities. When he revisited the river in 1882, Twain called the river "dead beyond resurrection."[8] On the lower Elkhorn, in Dodge, Washington, Douglas, and Sarpy counties, the Elkhorn River Drainage District and the Elkhorn Valley Drainage Districts completed a number of similar straightening projects prior to 1912. The projects improved drainage in the valley and contributed to minimizing flooding, but they worked only up to a point. The severe weather characteristic of the Great Plains produced snowmelt, ice jams, and thunderstorms that could overwhelm the best efforts by drainage districts to perfect the "flaws" in the Elkhorn Basin.[9]

Sometime around 1910 the Western Sarpy County Drainage District dug a series of ditches to drain bottomlands just east and upstream from the Elkhorn's mouth. Although this area normally lies outside the Elkhorn basin and drains east, ultimately into the Platte, Elkhorn floodwaters often washed into the area.[10] Also at about this time—sometime before 1911—several drainage districts modified Rawhide Creek, which flows through Colfax and Dodge counties, by constructing an extensive system of drainage ditches. When the drainage districts dissolved, however, the ditches deteriorated from lack of maintenance. As the ditches and the Rawhide Creek channel filled with silt and debris, they ceased to function properly and, in fact, caused an increase in the frequency of flooding experienced by adjacent lands.[11] However successful these projects were when properly maintained, their neglect pointed out the necessity of continual upkeep and the difficulty for small-scale, basically ad hoc methods of managing rivers and controlling floods.

The Flood of 1912

A severe flood in 1912 drove home those points. Conditions for flooding began taking shape in February that year when Nebraska saw its greatest monthly snowfall in at least thirty years. When that snow melted in March and April, it created massive runoff that washed out bridges in the valley and left old-timers debating whether it was worse than the flood of 1881. At Stanton, centrally located in the basin, seventy-six inches of snow fell during the winter of 1911–1912—more than double the thirty-inch average for the previous twenty years. March alone saw thirty-six inches of snow fall at Stanton. By March 25 ice still covered the Missouri River and nearby residents expected serious flooding as warm weather melted snow and the icy crust on the rivers. Railroad officials, fearing for the well-being of their lines, met at Omaha to plan for the expected flood. They ordered dynamite distributed to explode ice gorges in the river to facilitate drainage.[12]

Much like the river's flowing water, ice—solid water—could be both friend and foe to people in the Elkhorn Basin. Before artificial refrigeration came to the Elkhorn Valley, the river provided ice that residents could use to preserve food and for other cooling needs. Pickerel Lake, actually a river bayou near Ewing, was just the right size to work with a team of horses—when it was frozen. The team pulled a plow beam fitted with special knives across the ice. Men sawed the ice across the team's trail and pulled up 300-pound blocks to be stored for use in warm weather. Neligh photographer Emanuel Wolfe record-

Figure 3-5: Before artificial refrigeration came to the Elkhorn Basin, the Elkhorn and its tributaries provided ice to basin residents. Here, workers harvest ice near Neligh probably between 1905 and 1912. Courtesy of Nebraska State Historical Society.

ed a scene of men cutting ice, most likely from the Elkhorn River and probably sometime early in the twentieth century (see Figure 3-5). A mill pond, such as the one created by the mill dam at Pierce, not only provided power for the mill but also created an ideal source for harvesting ice in the winter. At Pierce, this created a spin-off industry, which included large storage buildings just north of the mill pond. In the summer, Pierce's ice works supplied merchants and residents locally and shipped carloads of ice to other towns. Residents harvested ice from the Elkhorn and its tributaries at least as late as 1912, when the *Norfolk Daily News* reported that ice harvesting near the city had left open a channel that allowed the river to flow freely and that had helped prevent ice gorges from damming the heavy flow created by the massive snowmelt that year. The railroaders' ice-blasting plan wouldn't be needed at Norfolk, but the absence of complicating ice jams wasn't enough to prevent flooding.[13]

The expected flood materialized at Norfolk at 5 p.m. on Wednesday, March 27, when the North Fork, an Elkhorn tributary that joins the main river southeast of Norfolk, rose high enough to halt operations at the town's mill. As has been the case historically, Norfolk's flood came as the North Fork overflowed

Figure 3-6: The 1912 flood was fairly shallow in Norfolk's streets, but buildings with basements, such as the one immediately behind the seated figure, took on several feet of water. Courtesy of Elkhorn Valley Museum.

north of town, flowed through town, and rejoined its channel in the southeast part of town.[14] By 7 a.m. Thursday, flood warnings arrived from Pierce, a North Fork community upstream from Norfolk. Two hours later water covered the southeast section of the city, and by 10 a.m. people there were abandoning their homes for high ground. The North Fork continued rising until Friday when floodwaters crested eight to ten inches higher than the old mark recorded above the mill dam. Flood watchers congregated on the corner of Norfolk Avenue and Fourth Street and used chalk to mark the height of the historic flood.[15] Two men helped dramatize the flood for posterity by posing on a flooded street for a photographer. One rode a bicycle through water deeper than the wheel rims as another man sat calmly smoking a pipe on a chair placed in ankle-deep water in the middle of the street (see Figure 3-6). As quickly as the flood materialized, it dissipated. At Clearwater, upstream from Neligh, the Elkhorn dropped four feet during the night of April 1–2.[16]

High water damaged structures in downtown Norfolk, washed out roads and railroads, and inundated farmland surrounding the city. Water undermined pavement in town and made jagged rollercoasters of sidewalks. Flooding ruined goods stored in basements of downtown businesses. At Norfolk's Sugar City Cereal Mill, perched as one would expect on the bank of the Northfork, high water—too much of a good thing—brought the hydro-powered mill to a standstill and erosion collapsed part of the flooring. Madison's mill, too, ground to a halt as the flood washed out its dam and led the owner to consider converting to electric power. The newspaper's presses stood in about five feet of water. Washed-out railroads crippled train traffic to the south and east. Without freight service, Norfolk's stores experienced shortages of meats, which sold out in several stores by Saturday night. Southwest of Norfolk, the Elkhorn swelled to three-quarters-of-a-mile wide and swamped farms. In one case it flooded a corral with as much as four feet of water. The strong current carved a hole, seven feet deep in places, through the corral. Early estimates placed the cost of damage in and around Norfolk at $25,000.[17]

A washed-out bridge at Wisner crippled wagon traffic and mail delivery between the town and the southern part of Cuming County. The images of Neligh Mill and the bridge adjacent to it drive home the importance of transportation in the basin for the movement of people and goods. The Elkhorn drove local economies as it turned milling equipment, but it could also stifle productivity when it washed out bridges and closed roads. In this flood it also left about thirty telephone wires broken and tangled in the river, cutting telecommunications between the same regions. A temporary bridge soon restored traffic into town, but town officials reported that a new bridge would not be needed.[18]

By the time of the 1912 flood, Norfolk's residents had clearly devoted considerable thought and action to flood control. Twelve families in southeast Norfolk began collecting boats and wagons to seek higher ground within five minutes of a blast from the town's fire whistle, which doubled as a flood warning. As waters rose on Thursday, laborers built a coffer around the town's waterworks to keep its machinery dry. They succeeded and the town enjoyed uninterrupted water service throughout the flood. The sewer system didn't fare as well. Engineered to drain water and waste from the town's buildings, the sewer backed up, which caused a number of basements to flood. With the sewer system disabled, waste water from the electric plant could not be disposed of, forcing the plant to close and to cease supplying electricity to the city.

Reports of residents observing the rising waters from a dike and of waters streaming over a retaining wall indicate that the town had built structures to defend itself from flooding. Flood warnings from upstream gave entrepreneurs most of the day Thursday to move inventory from flood-prone basements, thus averting costly losses. In other defensive or adaptive responses, Norfolk's residents turned to footwear and watercraft. One shoe dealer sold out a $200 stock of boots Friday morning, restocked, and sold out again before noon. Businesses that normally delivered goods to customers by wagon resorted to boats as delivery vehicles.[19]

Waters from tributaries further swelled the Elkhorn. From the courthouse tower in West Point, the valley looked like a sea with islands of high land dotting the waterscape. The flood rendered all of Dodge County's "important" bridges inoperable, an estimated $125,000 loss. Neighboring Cuming County also lost a number of small bridges and culverts, but the *Republican* claimed they were not significant losses. Waters washed away the approaches to the river bridge at West Point, but, again, the paper reported the damage as "not heavy." West Point lost its mill dam, a significant blow because the mill powered the pump for the town's waterworks.[20] At first the city seemed destined to replace hydro-powered pumps with steam pumps at an annual cost of $2,000 to $4,000. The Electric Light Company, however, offered to sell the necessary power to the city for about $1,600 annually—an offer the city's leaders accepted. The cost of a motor, to be supplied at the city's expense, added to the flood's costs to West Point.[21] Here, as would happen soon in Neligh, Elkhorn Basin residents replaced the river's power with energy derived from fossil fuel. In this stage of modernization and progress, the river was losing its usefulness.

The *Republican*'s report underscored the significance of humans changing the basin's face. The paper argued that while the 1881 floodwaters likely reached farther into town, there were fewer sidewalks and other embankments to impede its progress.[22] This, however, contradicts modern interpretations, which argue that structures in the floodway increase the severity of flooding.[23] Such misconceptions of flood mechanics mark human responses through much of history.

Near Logan Creek, which flows south from Cedar Creek to its confluence with the Elkhorn near Winslow and which is among the Elkhorn's largest tributaries, flooding washed out railroad tracks disrupting Bancroft's rail service for more than two days. Some, perhaps much, of the snow-melt escaped Logan bottom via a drainage ditch. By 1912 humans obviously had taken action to minimize the effects of flooding.[24] The damage reports from the 1912 flood also point out the increasing complexity of human life in the Elkhorn Basin. The 1823 flood recorded by the Duke of Württemburg inundated seasonally vacant Indian dwellings and did little or no damage to Indian crops. The 1912 flood damaged bridges, roads, railroads, houses, businesses, telephone lines, waterworks, and electricity-generating plants. By placing things of value in the floodplain and hitching local economies to the river's motive horse power, humans motivated themselves to try to control the river and also left themselves more vulnerable to its mercurial nature. This context helps us understand how floods, a natural, value-void occurrence, came to be seen as damaging, and rivers, when anthropomorphized, could be considered wrathful. It also helps us to understand that whatever damage the Elkhorn and its tributaries inflicted really cannot be considered natural disaster. People chose a close relationship to the Elkhorn and its floodplain, and those flood damages deserve to be called man-made.

Downstream, massive floods in the Mississippi Valley in 1912 and 1913 brought pressure from politicians and floodplain landowners for the federal government to do something to control future floods. President Woodrow Wilson ordered the Mississippi River Commission, which Congress had created in 1879 to study flooding and the effects of a broken chain of levees built by local levee districts, to submit a comprehensive report on flooding and flood control. The commission's report considered six kinds of flood control and recommended the levees-only approach that had held sway for decades.[25]

The levees-only approach to controlling the Mississippi drew criticism from abroad. Sir William Willcocks, a British engineer who had designed the Aswan Dam for the Nile and had helped regulate the Tigris and Euphrates rivers, found the American approach to the Mississippi inadequate and even irrational. Willcocks explained his critique of levees only in an interview with a *New York Times* reporter:

> [Y]ou have allowed the levees to be put up quite irregularly, and as each upper levee has been erected floods have worried those down below. . . . You have deprived your river floods of the opportunity of spreading out at the sides by contracting them, and so, perforce, they have risen into the air—risen frequently, until, with dire results, they have overflowed the natural and artificial barriers which normally confine them. . . . Therefore it seems to me that you really have brought the greater portion of your vast flood problem upon yourselves by bad management.[26]

For the Mississippi, Willcocks recommended that the St. Francis Basin, three million acres of forest land, be protected from development so that it could serve as a natural flood control reservoir. Furthermore, he recommended that the federal government assume a centralized, comprehensive approach to managing the river, and that affected states be required to follow it. Among other things, the centralized approach must ban construction of levees in unsuitable places. As Willcocks saw it, every human impediment to floodwaters "disorganizes the river and hurts navigation." Congress and the Corps of Engineers ignored these critiques and stuck with levees only. Alternative approaches cost too much and limited floodplain development.[27] Some of Willcocks's ideas would take hold in the United States, but not for decades.

Despite the progress inflicted upon the Elkhorn landscape, floods continued, flood damage continued to mount, and flood fighters assumed an even more aggressive strategy. Around 1914, workers dredged the Elkhorn's final twenty miles, creating a channel combining segments of old and new channels. The length and gradient of the natural channel can no longer be determined exactly. They also created an entirely new channel for nearly all of Logan Creek, which flows nearly due south from the vicinity of the Cedar-Dixon county line to its mouth near Winslow in eastern central Dodge County. Work began sometime before 1917 and continued sporadically until completion in about 1926. The new Logan Creek ran about 70 miles, less than half of the creek's original 150-mile course. Several drainage districts participated in the project under a well-coordinated plan that brought relief through reduced severity and frequency of flooding. Then again, the straightening project effectively moved the creek about a mile from the Bancroft mill that depended on the stream for power. The mill's owners compensated for the rechanneling by enlarging the mill dam to capture more runoff from rainfall. The scheme didn't work and the mill closed in the early 1920s. According to the USACE, this intensive, comprehensive, offensive approach to manipulating a waterway paid dividends in the form of reduced flood damages.[28]

The Floods of 1917

The Elkhorn's main channel, however, received no such treatment and soon wrought more destruction on human assets placed in its floodplain. Floods hit Norfolk twice in 1917. The first, in March, came when runoff from melting snow sent streams out of their channels, onto their floodplains, and brought floodwaters to their highest mark yet in Norfolk. The swollen Northfork threatened to rip the M. & O. railroad bridge from its moorings as it pressed ice against the structure. Unlike 1912, when ice harvesting had cleared a channel for the river, workers used dynamite to break up the ice. The bridge, which had moved several inches off-line, settled back into place after the pressure was removed. As had happened in 1912, water flooded the *Norfolk Daily News* building and stopped production of the paper for several days. At Norfolk the river expanded to nearly a mile across and flooded the eastern and southeastern part of town. Police evacuated families stricken with terror by the sound of floodwater rushing into and through their homes. On a lighter note, the high water provided sport for some automobile owners who braved the flooded streets for amusement, perhaps because of the novelty from submerged tailpipes spitting exhaust gases through the water and making the cars seem like bubbling motorboats. Some motorists ran sightseers or evacuees through the high water on their cars' running boards. Some got stuck when their cars stalled. Other residents fought the flood by stacking sandbags along the railroad embankments to raise and strengthen a de facto dike—and to protect a vital transportation corridor (see Figure 3-7). The March flood, which surrounded a number of homes, disrupted commerce, and caused three deaths, led Norfolk's city council to assume the mantle of responsibility for flood control in town and, according to a newspaper report, begin developing new flood control plans.[29]

Two months later, heavy rains in late May caused widespread flooding, including rare overflows on the upper Elkhorn as far upstream as O'Neill. Reports of high water on the North Fork at Pierce reached Norfolk on Saturday, May 26, prompting the mill operators at Norfolk to open wide the gates of the mill dam to allow the maximum amount of water to escape downstream. In their dance with the river, the mill keepers recognized in this moment that the Elkhorn should lead. To save their equipment and with the hope of minimizing damage to nearby property, they let it. In O'Neill the Elkhorn washed out one bridge and submerged another. Reports from Inman claimed the Elkhorn had spread to four miles wide.

Further downstream at Neligh, the Elkhorn rose to record heights and breached the dike between the upper and lower mill dams in two places, each 100 feet wide. South of town the river covered roads and flooded Riverside Park located just across the river. The surging Elkhorn carved a new channel, which left the water-powered mill high and dry. During World War I, the mill

Figure 3-7: On March 23, 1917, workers, including this unidentified individual, stacked sandbags on top of railroad grades in an effort to keep flood water out of downtown Norfolk and to protect the rail service. This photograph inspired the book's title. Courtesy of Elkhorn Valley Museum.

relied exclusively on a diesel engine. After the war, the mill built a new flume (the channel that carried water from the river to the machinery that converted moving water into power) and penstock (water intake structure) and brought the Elkhorn back to its previous channel to restore water power, which worked until the flood of 1920.[30] At Neligh, the Elkhorn not only damaged the milling operation, it effectively abandoned the mill. But people refused to accept that result and worked the river by moving its channel so the river would work for them. Perhaps, too, they wished to preserve the boating and fishing opportunities afforded by a dammed Elkhorn, captured in a period photograph (see Figure 3-8).

By Tuesday, May 29, the flood's crest reached the Norfolk vicinity where farmers west and upstream of town awoke to find low-lying pastures covered with water. Boys in the area made the best of the flood by scooping up fish that had been washed into the pastures. By this time the flood danger on the North Fork had passed and the Elkhorn's rise appeared to have slowed.[31] That appearance proved deceiving, however, as the river continued to rise at Norfolk through Wednesday morning thanks to additional rain. Although reports

Figure 3-8: Basin residents valued the Elkhorn into the twentieth century not only for the work it could do, but also for the recreation it afforded. Here, in about 1910 a man fishes upstream from the Neligh Mill dam, which had effectively turned the Elkhorn into a mill pond. The mill and the bridge over the Elkhorn can be seen in the background just above the boat. Courtesy of Nebraska State Historical Society.

indicated that the Elkhorn had reached lands never before affected by floods and covered considerable farmland west of Norfolk, the river had not risen high enough to disrupt railroad traffic.[32]

More rainstorms over the next week kept the Elkhorn high. At Norfolk, the North Fork first receded and then climbed again before beginning a slow decline that lasted into the second week of June. At Pilger, the swollen Elkhorn backed up Humbug Creek, which flooded the town when runoff from a 1.8-inch rainfall during the night of June 3–4 took the creek over its banks. Even farther downstream, a local cloudburst near the Elkhorn's mouth at Gretna on June 5 caused flooding there. The Elkhorn inundated hundreds of acres and crop damage estimates climbed into thousands of dollars.[33]

The Elkhorn did not flood in isolation in 1917; enough other rivers flooded to inspire the Ransdell-Humphreys Flood Control Act of 1917, which marked the first time the federal government authorized levee construction with no direct benefit to navigation. At the levee, though, not much changed and local entities continued to do most of the work of construction and maintenance.

Most people thought land-use planning and flood control were matters for local control and many questioned whether it was the federal government's business to protect private property from flooding.[34] Attitudes were changing, but slowly.

The Flood of 1920

Elkhorn Basin residents enjoyed a reprieve from major flooding until several feet of snow and two weeks of April rain turned the Elkhorn into a machine of destruction in the spring of 1920. A headline in the *West Point Republican* lamented, "Waters Cover the Earth."[35] Water didn't cover all the earth, but it reached record heights at Inman—about fifteen miles downstream from O'Neill—on April 20 making roads impassable for miles. The Elkhorn swelled to two miles wide at Ewing. At Neligh, the Elkhorn rose eighteen inches during the twenty-four hours between April 19 and 20. At nearly midnight, the Elkhorn destroyed the impoundment dam at the Neligh mill, causing several thousand dollars in damage. As it had a few years earlier, the Elkhorn migrated away from the mill leaving the machinery silent. The mill's owners finally gave up on the Elkhorn. From then on, diesel-generated electricity would grind the mill's grain into flour.[36]

Downstream, West Point received five inches of rain and about a foot of snow during April's final two weeks. The *Republican* proclaimed 1920 to be the worst flood year since 1891. The flood damaged several bridges, roads, sidewalks, and bottomland crops and delayed trains and mail delivery. Unlike previous floods, however, the spring 1920 flood involved channel modifications made the previous winter. Modifications near the Elkhorn bridge at West Point diverted water in such a way as to endanger the bridge's east end. Workers used dynamite to erase the modifications from the channel.[37] Basin residents continued to learn on the job when it came to dealing with the unpredictable Elkhorn's destructive tendencies. It also reflected a piecemeal approach to modifying the river's channel that showed limited understanding of the implications of altering the floodplain and the limits of well-meaning humans' efforts to impose progress on the landscape.

Flood stories still dominated the front page of the *Norfolk Daily News* on April 26 with reports of high water destabilizing railroad bridges. Already weakened to the point that train traffic across them had been suspended, railroad bridges west of Norfolk could no longer sustain even pedestrian traffic by the 26th. Until that time, trains apparently stopped at the bridge while passengers disembarked, walked across the bridge, and boarded another train. If the Elkhorn's swirling waters were not hazardous enough, uprooted trees and other debris carried by the flood's swift current slammed into pilings further weakening the structures. The flood also disrupted Norfolk's telecommunications as the swollen Elkhorn claimed at least two telephone poles.[38] The water power that drove milling equipment to create value could also wreck infrastructure at mills and bridges, imposing a clear cost to doing business with the river and being tied to its floodplain.

The Norfolk newspaper speculated in its reports of April 26 that the flood's crest had passed Norfolk and that downstream communities could soon expect to be inundated. Reports from Beemer and West Point that day both confirmed that flooding was already underway in those communities. At West Point, the water reached its highest point since 1888, rushed down at least one of the town's streets, covered bottomland on the town's west end, and damaged surrounding farmland.[39]

In a rare occurrence, the 1920 flood claimed a human life. The Elkhorn washed out the approach to the bridge at Tilden. To ferry people and supplies across the river, residents had strung a cable—with a boat attached—across the river. C. A. Davis, who lived seven miles north of Tilden, and another man tried to reach Tilden using the ferry, but the boat capsized sending both into the water. Davis's companion swam to safety, but the cold current swept Davis to his death.[40]

At Norfolk, high water moved into a new channel, which left thirty people on a 160-acre island nestled between the new channel and water flowing through the old channel. Boats brought supplies from the mainland as the islanders ran low on food and vital supplies, even down to matches. City officials investigating the situation proposed establishing regular boat service to the island families. According to a *Norfolk Daily News* report, the new channel cut across a cornfield from which it had been diverted by a dike.[41] Assuming the report's veracity, human interference with the river kept it from its natural course and apparently lulled inhabitants into a false sense of security, which led them to build in an area rightfully belonging to the river—an area the river reclaimed during the 1920 flood.

Responses to the flood included an effort to reinforce bridges on the Harvest Trail, a north–south highway connecting Butte and Red Cloud, Nebraska, against the effects of future floods. In Holt County, a new steel and concrete–

floor bridge went up across the Elkhorn south of O'Neill and a bridge built to span an overflow channel for the river. Finally, workers raised the road grade by two feet through the river bottoms. Similar bridges spanned the South Fork and Dry Creek, two Elkhorn tributaries also crossed by the trail.[42] Elkhorn Valley residents learned that they needed stronger and higher infrastructure to coexist with the river.

Weeks later, in June 1920, landowners responded to the flood by proposing an Elkhorn drainage district to undertake channel modifications and other measures to control the river. Opponents of the project wrote to the *West Point Republican* that the project was too costly and that the unstable nature of the riverbed's soil made the project impractical. "It is generally known that a fallen tree in the stream is enough to so divert its flow as to form a new and different channel at that point, of such unstable character and nature are its banks and the lands tributary," they wrote. They also argued that, "It does not logically follow that because of the unusually high water of this season, the highest of many years, that the same will recur every year hereafter."[43] They were right, of course, but it should have been clear even at this point that flooding was a customary part of living with the Elkhorn. Furthermore, for sixty years people had placed more and more assets in the floodplain. The river didn't have to work as hard or flood as high to inflict more damage. Within twenty years, a series of even worse floods might have left those drainage district opponents wishing they had supported the project.

The "Dry" Decades

Somehow the Elkhorn seems to have passed most of the 1920s and 1930s with little to no significant flooding. High water surely visited the basin in those years—the Elkhorn Valley Museum displays a photo of water covering Norfolk Avenue in 1936, for example—but evidence of major floods the likes of which are chronicled in most of this book's stories is scarce. A catalog of the state's floods compiled by the Nebraska Department of Natural Resources lists no events in the Elkhorn Basin between 1920 and 1940.[44] On one hand, this might not seem odd given that drought and dust dominated life on the Plains for most of the 1930s. On the other hand, drought-dessicated soil in Nebraska's Republican River Basin contributed to extraordinary runoff that produced one of the state's deadliest and most costly floods, the Republican River flood of 1935.[45] Despite the apparent lack of remarkable flooding in the

Elkhorn Basin, though, people continued to manipulate the Elkhorn and attitudes about flooding and flood control continued to evolve.

In the 1920s and 1930s more localized modifications came to the river and the federal government expanded its influence in the basin. In 1924, Norfolk-area residents organized the Norfolk Sanitary District. It did not take the district's organizers long to act on modifying the North Fork of the Elkhorn in hopes of curbing its destructive floods. From the mill dam at Norfolk to the North Fork's confluence with the Elkhorn, the district straightened and cleaned the stream's channel and built several bridges across it.

The 1927 Mississippi River flood, one of the most calamitous to date, added to a growing number of people convinced that the federal government should expand its efforts to fight flooding. It led to a dramatic change in federal policy codified in the Flood Control Act of 1928, which gave the federal government (via the Corps of Engineers) responsibility for flood control in the Mississippi basin.[46] It would take years, though, for that notion and for federal influence to flow upstream from the mighty Mississippi as far as the Elkhorn.

In 1936 the Norfolk Sanitary District built a flood control structure at the mill dam, which allowed operators to control the flow of water, to a certain extent, past the obstruction. That same year Congress expanded federal authority yet again in the Flood Control Act of 1936, which extended federal authority over flood control out of the Mississippi basin and to the entire nation. The act labeled floods a "menace to national welfare" and gave the Corps of Engineers broad authority to fight the menace. The Corps could build any flood control project it wished, as long as Congress agreed and supplied the money. The act also provided a standard for making decisions on flood control projects: cost-benefit analysis. The benefits had to exceed the costs. The new federal approach would find its way to the basin, but not for nearly a decade. Meanwhile, the New Deal would place its mark on the Elkhorn. In 1940, Works Progress Administration laborers built a dam on the North Fork at the city of Pierce to form a small lake for recreational use and minor flood control.[47]

Further downstream, other communities undertook a variety of flood control measures hoping to protect themselves from the damages of high water. During the 1920s, Pilger-area workers dug a new channel for Humbug Creek and built a system of levees along the lower three miles of the creek to protect Pilger and nearby farmland. Pebble Creek, which flows southeasterly through southern Cuming and northern Dodge counties, lost several sharp bends (dates of the work are unclear) when humans cut through the banks to straighten the stream's course. Improvements also included a system of small levees. Because individual landowners made these modifications piecemeal, however, they were reportedly effective in combatting only minor floods.[48]

Conclusion

From the 1890s through the 1930s, much changed in the Elkhorn Basin. Electricity arrived to power modern machinery and labor-saving appliances. Initially, some of that electric power originated in the Elkhorn's flowing water. In a sense, early electrification brought the river's power beneficially to businesses, homes, and streets of places like Neligh. Maybe at that point the people of the basin and the river had their closest relationship. But that moment passed in a blink—or in the gurgling whoosh of high water. The unruly Elkhorn and its unstable, sandy banks made a shaky foundation for modernity. More reliable, predictable, and robust sources of energy elbowed the Elkhorn out of the picture after 1920.

The Elkhorn Basin boasted what was probably to that point its largest human population, which lived in structures and communities intended to be permanent, or at least very long term. These people had placed assets and infrastructure of unprecedented value and complexity within reach of the Elkhorn's floodwaters. To protect these assets, people took on the river in new ways and on a greater scale. They straightened it. They tried to contain it. But they didn't back away from it. They formed quasi-governmental agencies to manage it. They welcomed a federal agency to help. Cities and private landowners both tried to keep the river away from valuable roads, railroads, buildings, and crops. Despite their best efforts to perfect the once useful Elkhorn, it remained an imperfect river and high water kept coming.

Chapter 4

THE FLOODED FORTIES AND "THE WORST FLOOD IN HISTORY"

"Damage by Elkhorn Defies Estimation"

Headline in the *Fremont (NE) Guide and Tribune*, June 13, 1944

By the 1940s residents of the Elkhorn Basin had almost completely divorced themselves from their practical relationship with the river and its tributaries. Flowing water no longer powered flour mills and no longer generated electricity. One of the great attractions to the floodplain had become irrelevant, but the sons and daughters of the Elkhorn inherited consequences of decisions their mothers and fathers had made in an earlier, different age. Traffic still flowed on the railroads and roads linked intimately to the river. People had put considerable time and effort into protecting themselves and their assets from high water, yet they could not entirely stop or control high water, which continued to arrive in many springs and some summers and continued to impose a cost for attempts to conquer the basin. Modernization, in forms such as steam and electric power derived from fossil fuels rather than hydropower furnished by the Elkhorn, contributed to the distance between people and river, but settlements and agriculture in the floodplain continued to connect the two. The decades leading to the 1940s saw residents band together in relatively small-scale collective actions such as drainage districts to control or react to their unruly liquid neighbor. The 1930s brought increasing involvement from even larger cohorts of citizens extending outside the basin to include residents of the state and the nation via the work of state and federal government agencies. This reflected developments that can be traced to Progressive Era and New Deal policies and a changing ethos guiding the United States and its relationship to the environment. Despite wide-ranging attempts to curb flooding along the Elkhorn and its tributaries, the 1940s brought two of the most devastating floods in the basin's history. Severe flooding in several parts of the Missouri system, of which the Elkhorn is part, in 1944 would prompt unprecedented levels of federal management of the Great Plains rivers. The Elkhorn, at the relatively sparsely populated and developed end of the line, exposed the operation of the new regime and its limits.

The Flood of 1940

During the night of Monday, June 1, to Tuesday, June 2, 1940, cloudbursts dumped a foot or more of rain on parts of the basin. The Elkhorn rapidly came over its banks. Rain in the north-central basin swelled the North Fork, a tributary that flows from northern Pierce County, past the town of Pierce, and on to its confluence with the Elkhorn at Norfolk. But Norfolk's citizens did not fear a North Fork flood because of a spillway and a channel-straightening project built by a drainage board. The Elkhorn was a different matter. Just upstream from Norfolk, farmers reported a four-foot rise between 6 a.m. and 8 a.m. on June 2.[1] Although it was on the storm's fringe and received only 1.08 inches of rain, the Elkhorn made a similar rapid rise Tuesday morning at West Point. Witnesses considered it "unbelievable that such a catastrophe which completely inundated all property in less than a half hour could happen."[2]

At Pilger, high waters disrupted the waterworks system and electrical service, leaving the town without power for more than ten hours during the night of the storm.[3] Turbulent waters rushing through Humbug Creek to the Elkhorn churned through the earthen levees built during the 1920s.[4] Again, bridges suffered damage, which threw railroad schedules behind for a week or more after the flood.[5] Most damage to the Chicago and North Western railway lines came between Norfolk and West Point where the flood damaged a bridge, washed out several bridge approaches, and washed out about four miles of track, some of which it carried a quarter mile downstream.[6] By the next day the Chicago and North Western would report thirty-five miles of track washed out in the West Point-Stanton area and the Union Pacific reported closing its mainline traffic to Norfolk.[7] The washout left West Point, Beemer, Pilger, Wisner, and Stanton without rail service until the track could be repaired, a period of at least ten days.[8] The flood also disrupted telecommunications leaving residents of West Point and Pender to communicate via shortwave radio.[9]

The crest of the flood rolled down the valley like a wave. By Wednesday morning the crest had passed Stanton allowing residents there to begin assessing the damage exposed by the receding water. Mud and silt buried the corn crop, which would have to be replanted as soon as the soil dried, and covered pastures and grain fields. Highway traffic returned nearly to normal around Norfolk by Wednesday afternoon, but washouts left some sections closed. The Nebraska State Highway Department warned motorists using Highway 275, the main route between Norfolk and Omaha, to beware of flooding as the crest moved toward West Point.[10] When the flood swept over the highway, it deposited trees and other debris, which effectively blocked the highway from 1 a.m. to 4 a.m. on Thursday. By 8:30 a.m. traffic faced only minor obstructions between West Point and Norfolk.[11]

Even before the flood crested in West Point, rising water forced families to evacuate their homes or to seek assistance.[12] Some found food and shelter in a local church that had been converted into the Red Cross relief headquarters. Some flood refugees also slept there until it was safe to return to their homes.[13] The hatchery at West Point lost more than 1,200 chickens and two tons of feed in the flood. Water in parts of the city reached higher levels than marked in the 1912 flood and the city lacked mail service from Monday until at least Thursday.[14]

Despite these damages, West Point went on with its eightieth birthday celebration while under warning that the flood's crest, which would peak at thirteen feet above normal, was expected to arrive sometime Thursday during the celebration. *The Norfolk Daily News* even reported that "no great trouble is anticipated" from the crest.[15]

Just as residents of the upper basin must have thought their flood troubles were ending on Wednesday and Thursday, a second rainstorm hit Norfolk and areas east and south of Pilger. By this time counties near the Elkhorn's mouth began to feel the effects of the heavy rains. At Winslow and Nickerson, where Logan and Maple Creeks, respectively, empty into the Elkhorn, floodwater covered farmland. In Dodge County, Logan Creek and the Elkhorn overran their banks to flood an estimated sixty farms. Arlington, the Washington County seat, reported its highest flood since the ice gorge of 1912. In Douglas County, the Elkhorn topped its banks in three places near Gretna on Thursday morning.[16]

The renewed rains and flooding compounded problems with the basin's already crippled transportation system. The Chicago and North Western, forced to reroute its Black Hills traffic from the washed-out West Point–Norfolk section to its Oakdale branch line, suffered a washout and wreck six miles on the branch line early in the morning of June 6. Four people were injured as a locomotive overturned at the washout and took a baggage car with it.[17] As late as Saturday, June 8, reports from Scribner indicated the Elkhorn was still rising.[18]

Responses to the 1940 flood, arguably the worst to date, ranked among the most comprehensive to date and included action by local, state, and federal governments. Representatives from the University of Nebraska's agricultural college visited flood-worn northeast Nebraska to assist county extension agents in advising farmers how to limit their crop losses. Their suggestions centered on replanting flood-damaged corn, sorghum, and forage crops with early-maturing varieties.[19] Replanting crops, however, required the Agricultural Adjustment Administration (AAA) to modify its regulations governing crop allotments.[20] The Red Cross, in addition to the aforementioned services, provided serum for typhus vaccinations and supervised the disposal of lime and other fumigating agents used on the town after the flood receded.[21]

Throughout the basin, road crews set about repairing roads and bridges. Cuming County alone suffered an estimated $200,000 (1940 dollars) in damage to public facilities, more than $100,000 of which came from damage to roads and bridges.[22] Responding to the destruction, the Board of Supervisors of Cuming County called an emergency meeting on June 7 to explore sources of federal aid. The board decided to request $100,000 to offset repairs to bridges and roads.[23]

Within two weeks of the flood, Works Progress Administration (WPA) laborers repaired twenty-five bridges in Cuming County. In more unsavory work, they removed about eighty-five tons of debris from West Point's streets and buried carcasses of drowned animals.[24] Railroad workers took two weeks to make temporary repairs to the Chicago and North Western line that allowed the railway to function while permanent repairs proceeded.[25] On July 2, the county supervisors spent $5,463 from the Emergency Bridge fund for further repairs.[26] By August it had become apparent that the damage to Cuming County's bridges would exceed the amount budgeted for repairs and the supervisors raised the Emergency Bridge fund levy for 1940 from .0125 mills to .224 mills—nearly an eighteen-fold increase.[27] Although the supervisors dropped the levy the following year, they left it at .13 mills rather than return it to the pre-flood rate, perhaps indicating they had learned a lesson from the 1940 flood, or they were still paying for its damages, or both.[28]

One 1940 flood response foreshadowed responses to the next serious floods. A representative of the US Department of Agriculture (USDA) visited the West Point Community Club's regular July meeting to gather information about damages incurred in the flood. The information was turned over to government engineers to help them formulate new flood control plans. The engineers, however, apparently did not use the information because a similar meeting four years later collected the same type of information for the same purpose and found its way into a report by the USACE.[29]

Through the first ninety years of non-Indian settlement in the Elkhorn Basin, floods assumed increasing importance to its residents and elicited increasingly complex responses. In 1857, high waters merely impeded West Point's founders in their travels to the town site. Early settlers had little choice other than to bear the flood's frustrations and wait for the water to recede. But as soon as they, and other town founders, built in the valley, especially in the floodplain, floods became increasingly problematic. The river did what it had always done, occasionally flow over its banks and onto its floodplain. The actions of settlers and town builders, though, turned the Elkhorn's floods into "natural disasters." Accordingly, humans began taking steps to safeguard their valuables, mainly the buildings that made up their towns and the transportation and communication infrastructure that had been built along the river. Initially, these efforts assumed a defensive character—dikes, levees, and floodways to contain rising water or divert it from valuable assets such as buildings and farmland. Over time, flood protection efforts assumed an offensive character—altering the river, usually through straightening or othwerwise "impoving" its channel, to prevent flooding in a particular locale.

Examining the response pattern to floods casts some light on human nature and American society, politics, and economics. Settlers who came to the Elkhorn in the late 1800s needed the river's power to succeed. But in its natural state the Elkhorn couldn't help them much. For American civilization to flourish in the basin, people had to improve upon or perfect the river. They built dams to raise a head to power mills, for example. They learned quickly, though, that this intimate connection to the river brought high costs as well as benefits. They built levees and dikes and straightened channels as part of a second campaign to improve the river, this time with respect to flood control. These efforts reflected the environmental ethos of the Progressive Era. Humans perfected the landscape and tried to control the river according to scientific principles and increasingly through government or quasi-governmental agencies. Reactions to the 1940 flood point to an increasingly sophisticated social and political system operating in the basin. Agencies of the federal government, which had expanded its reach through New Deal programs created during the 1930s, became part of flood response. The AAA, WPA, USDA, and

USACE joined local agencies and organizations in shaping how basin residents lived with the Elkhorn.

The 1940 flood clearly showed Elkhorn Basin residents their continued vulnerability to flooding despite the various defensive and offensive responses to that date. Yet hardly any residents attended a hearing to assess flood damages and to discuss projects to prevent or minimize future floods. Presumably little, if anything, was done in response to the 1940 flood. Yet a scant four years later, nature and the Elkhorn would twice send floodwaters onto the floodplain and these floods would reach higher and destroy more human assets than any before. These floods would draw serious attention and mark another turning point in the relationship between the Elkhorn and its people.

The Floods of 1944: Worst Ever

April showers made the spring of 1944 an especially wet one. That month saw the beginning of a thirty-day rainy spell that saturated soils throughout much of the region. Three years had passed since the destructive flood of 1940. The sons and daughters of the Elkhorn seemed to have put wet basements, flooded towns, and washed-out crops out of mind. Residents had done little, if anything, to change their relationship with the river and its tributaries since the inundation of 1940. A congressional report issued in 1934 deemed the Elkhorn unworthy of federal flood control measures because it was neither navigable nor an important producer of hydroelectric power. Flooding in the basins of the Platte and its tributaries never amounted to much, according to the report. Given that background and the drought of the 1930s, 1944's rainy spring probably had residents preparing for a typical, but unremarkable, Elkhorn flood. If that was the case, they were grossly unprepared when the runoff from two severe rainstorms, exacerbated by the wet spring and saturated soils, produced a pair of floods—one in the upper basin, one in the lower basin, both the worst ever—that would prod the government to change its stereotype of the region and change how the basin's residents interacted with the river.[30] The massive scale of 1944's floods and the dramatic changes they brought in the human dialogue with the river demand a close examination.

On Wednesday, May 10, rain clouds over the west-central Elkhorn Basin dimmed the afternoon sun. The massive storm covered large areas of eastern Nebraska and parts of Iowa, South Dakota, and Minnesota. At around 7 p.m., the rains came—hard. The cloudburst soaked all of Pierce County and portions of Holt, Antelope, Boone, Knox, Madison, Cedar, and Wayne counties for nearly two-and-a-half hours. The weather sequence repeated on Thursday, with rains drenching the same counties from 8 p.m. until about 10 p.m. At the storm's heart in Pierce County, aggregate precipitation reached ten inches.

Unable to hold the deluge, soaked soils sent much of the rain overland in an abnormally high runoff towards streams and rivers. All streams draining the area registered high stages after the storm. The North Fork flooded. Runoff swelled the Elkhorn, too, but most flooding occurred upstream from Stanton. High water covered about seventy-eight thousand acres along the Elkhorn. Tributaries flooded even more land. Far upstream, usually flood-free Atkinson flooded.[31]

At Neligh, normally considered the upstream limit of the basin's flood zone, the city's location on high ground on the Elkhorn's northeast side kept it from harm. Lowlands to the west did not fare as well when the river overflowed its banks on that side. Downstream, the Elkhorn flooded Oakdale, Tilden, and Meadow Grove. Pigs, hogs, and sheep drowned in the flood. Crops fared better. Because the waters receded fairly quickly—within two days—the flood damaged limited amounts of the small grains and alfalfa planted on bottomland.[32]

The North Fork flooded the town of Pierce to the tune of about $47,150 in damages. Much of that, about $36,000 worth, was done to 120 homes. The next largest dollar amount of damage, $5,000, affected the sewer system.[33] After draining Pierce County, located at the heart of the storms, the North Fork brought runoff to Madison County and the City of Norfolk by Friday, May 12. At 1:15 a.m., a shrill blast from Norfolk's fire whistle cut the humid early morning air as it called volunteers to haul sandbags to reinforce the town's dike near Braasch Avenue. About one hundred townsfolk shook off grogginess to bolster the dike. Even with the flood gates at the mill dam wide open, their work held for only hours before giving way to the swollen North Fork[34] (see Figure 4-1).

At about 2:30 a.m., a two-and-a-half-foot–high wall of water from the North Fork broke over the banks and headed for a Norfolk dance hall. The flood wave struck with enough force to wrench the 144-foot by 87-foot building from its foundation and carry it intact to the bridge over the North Fork at Third Street and Elm Avenue. According to the owner, previous floods had risen slowly and held to the river's course. This flood's fast, hard rise made it especially shocking to those who saw it and survived, even those who had survived other floods in the area. The flood also washed out five highway bridges between Norfolk and Oakdale by Saturday the 13th.[35]

Figure 4-1: The flood of May 12, 1944, left cars on Norfolk Avenue axle-deep in water. The basement windows of the *Norfolk Daily News* building (right center) suggest buildings with basements took on several feet of water.
Courtesy of Nebraska State Historical Society.

The stream climbed high enough to flood an estimated 861 acres in and around Norfolk to depths reaching four feet.[36] City engineer H. H. Tracy reported that the water reached two to three feet higher than the previous high-water mark in the city, and pilots who surveyed the scene from the sky described the Norfolk area as a giant lake.[37] G. L. Carlson, a longtime Norfolk resident, confirmed Tracy's report, telling the local newspaper that the high-water mark left by the 1944 flood on his 122 Norfolk Avenue building surpassed the mark of the 1912 flood, the city's previous worst, by two feet, four inches. A photograph of downtown Norfolk made on May 12 shows adults standing in nearly knee-deep water on a sidewalk and someone paddling a kayak down the middle of the street[38] (see Figure 4-2).

As the North Fork flooded the city from the north and east, the Elkhorn rose from its banks to flood the city's south side. Cresting a foot-and-a half above flood stage late on Friday the 12th or early on Saturday the 13th, the Elkhorn drove people from their south-side homes as it reached as far west as Second Street.[39]

Receding waters revealed the flood's handiwork. The business district and a large residential area—176 city blocks total—felt the waters of either the Elk-

Figure 4-2: The flood of May 12, 1944, made it possible to travel over Norfolk's city streets by kayak. Courtesy of Elkhorn Valley Museum.

horn or the North Fork. High water damaged 177 businesses and 459 homes and forced between 300 and 400 people to leave their homes, about 100 of them by boat. The flood interrupted telephone, power, and railroad services, and left a heavy deposit of mud and slime. Table 4-1 shows how the $1,236,000 of direct damages affected Norfolk's businesses, residents, and local government. These amounts do not include indirect damages such as losses incurred from interrupted commercial activities.[40]

Table 4-1: Dollar Cost of Flooding, Norfolk, May 1944

Businesses	1,006,000
Private homes, personal property	147,000
Municipal property	68,500
Flood fighting, rescue work (Red Cross)	4,500
Flood fighting, rescue work (others)	10,000
Total	1,236,000

Source: U.S. Engineer Office, *Interim Report on Flood Control for the City of Norfolk, Elkhorn River Basin, Nebraska* (Omaha: U.S. Engineer Office, 1946), 15.

Commercial losses ranged from disrupted rail service to lost inventory. Grocers suffered from an embargo on the sale of any goods touched by floodwaters—an effort to prevent an epidemic of flood-born diseases. Vegetables, such as lettuce, normally eaten raw, were to be destroyed. So, too, were canned and bottled goods. Authorities filed charges on Monday the 15th against a merchant who sold contaminated bottled and canned goods. Other foods, such as corn or beans, that would be boiled before being eaten were exempt from the embargo. Flooded merchants who sold goods other than food lost business as they closed for as many as ten days to clean debris from their stores.

Businessmen Ben Skalowsky and Pete Davis probably experienced more damage than the typical flooded Norfolk merchant. Skalowsky owned a twelve-year-old building on the 200 block of Norfolk Avenue, which housed his heating and air-conditioning business and Davis's Virginia Sandwich Shop. When floodwaters weakened the building, the wall separating Skalowsky's shop from Davis's restaurant collapsed, as did the floor in the building's rear, sending booths, restaurant equipment, and Skalowsky's truck crashing into the basement.[41]

Homeowners and merchants in the flood area found six to fifteen inches of mud on their floors and suffered flooded basements. Removing the mess and other debris was a labor-intensive effort hampered by the absence of those who were away fighting World War II. Local high school seniors ahead in their school lessons were allowed to leave school to work, for "high" hourly wages. A. W. Breyer, Norfolk's mayor, even considered trying to get prisoners of war to help clean the city. Breyer abandoned that solution when about fifty volunteers from Lindsay, Madison, Battle Creek, Tilden, and Columbus arrived with trucks on Sunday the 20th to help clear debris from Norfolk's streets.

State and local health officials added to the workload when they threatened to close shops with basements unless the owners removed all flooded goods, and then cleaned and disinfected the basements. Finally, all who relied upon a well in the flooded region for their water had to disinfect their wells to ensure potability of their water supply. The *Norfolk Daily News* helped the recovery by printing detailed instructions for well disinfection.[42]

According to news accounts and the U. S. Engineer Office, the May 1944 flood set the record for high water at Norfolk. By comparing peak flows measured in the North Fork at Norfolk during its largest floods up to and including 1944, Table 4-2 illustrates the severity of the 1944 flood. Flood watchers had been recording high-water marks at various spots around town for most of the twentieth century. Their measurements of the 1944 flood registered more than three feet higher than previous record heights.[43]

Table 4-2: North Fork Peak Flow during Notable Twentieth-Century Norfolk Floods

Flood Year	Discharge (cubic feet/second)
1912	12,900
1917	13,400
1944	23,400

Source: U.S. Engineer Office, *Interim Report on Flood Control for the City of Norfolk, Elkhorn River Basin, Nebraska* (Omaha: U.S. Engineer Office, 1946), 17.

These measurements allow comparison of high-water marks, but comparing damages from one flood to another is difficult. Only fragmentary evidence exists for most floods before 1940. Also, the increased level of development in the floodplain by 1944 gave that year's floods greater potential to do harm. Previous floods would have certainly caused more damage at 1944 development levels than they did in their times. By the 1940s, experts speculated that the existing flood hazard would curb future development unless protective works were built.[44] This again points out that high water in the Elkhorn Basin, a feature of living with the river that had to be expected but could not be precisely predicted, brought increasingly disastrous results thanks to human hands by 1944. These floods were disasters, but they weren't natural ones. And by the 1940s ideas about living in the floodplain and what it would mean for Norfolk's future were starting to change … again.

The May flood entered Stanton County Thursday and Friday and high water was receding by Saturday morning. Old-timers in the county reported the May 1944 flood to be about a foot higher than the previous high-water mark in Stanton County. By Monday, the flood had dropped far enough that the county's farmers could begin digging out of the mud deposited on their land, mending fences, and reassembling their scattered livestock. Some farmers had to move their animals several times to keep them on dry ground. Once those tasks were completed, however, farmers were left to wait for their land to dry enough that they could farm it.[45]

Although the May flood was most severe in the upper basin, lower-basin counties and towns felt its effects. The flood hit West Point on Sunday the 14th. Early that morning the Elkhorn began rising at about six inches per hour as it reached into the city park and Cuming County fairgrounds. City officials, aware of Norfolk's devastation, sounded the flood warning whistle at 11 a.m. Most of the water ended up on farms west of town, unlike the 1940 flood, which rolled into the city from the Plum Creek Valley east of the Elkhorn. In town, the river flooded gardens and basements and washed away an undetermined amount of property.

After a brief pause between 7 p.m. and 8 p.m., the river continued to rise at West Point until 11 p.m. on Sunday. Not until Monday did the flood's crest roll past Beemer. At its crest, the flood left the city park under five feet of water. It also disrupted traffic on Highway 275 and caused irregularities in rail traffic until Tuesday. A local weather observer reported that this flood never rose as high as the 1940 flood, missing the West Point flood record by eighteen inches. The volume and velocity of the crest had dissipated such that little trouble was expected further downstream at Scribner.[46]

Dodge County, too, felt the Elkhorn's rise. At Scribner, the Elkhorn had been rising since the afternoon of Friday the 12th. Between 7:30 a.m. and 8 p.m., the Elkhorn rose fifteen inches there to reach its crest.[47] The Elkhorn, although it overflowed its banks to flood ravines and low-lying areas near the river, caused damage described as "not being great."[48] All told, the Elkhorn flooded about 5,000 acres in Dodge County, 1,200 of which were planted in crops. Old riverbeds, timber, and pastures occupied the remaining 3,800 acres.[49]

One month, to the day, after the May storm hit, rains again drenched eastern Nebraska and parts of southeastern South Dakota, western Iowa, Kansas, and Minnesota. Between 10:30 p.m. on June 10 and 9:30 a.m. on June 11, a heavy thunderstorm centered over the headwaters of Maple and Pebble creeks in Cuming, Burt, Colfax, and Dodge counties dumped as much as fifteen inches of rain on the Elkhorn Basin.[50] Unofficial measurements made north of Howell registered sixteen-and-a-half inches of rain, of which fifteen-and-three-quarters inches fell in five hours. Within six hours, the average precipitation over 250 square miles at the storm's center was more than twelve inches.[51]

In a repeat of May's flood scenario, soils failed to absorb runoff from one of the region's most intense storms ever. The swollen Elkhorn covered an area about fifty miles long by twenty miles wide in the lower basin. It flooded roughly 84,000 acres extending from Stanton to Waterloo near the river's mouth including the towns of Stanton, Wisner, Beemer, West Point, Crowell, Hooper, Winslow, and Waterloo. Tributary flooding inundated an additional 61,000 acres, including the towns of Stanton, Pilger, Scribner, Dodge, Snyder, Leigh, Clarkson, Howell, Fremont, and Arlington. A gauging station at Waterloo recorded a peak discharge of 100,000 cubic feet per second—four times the station's previous record high. The surging Elkhorn raised the Platte River to flood stage from its confluence with the Elkhorn to the mouth of the Platte and added enough water there to a moderately high Missouri River to cause severe flooding from the Platte's mouth to St. Joseph, Missouri.[52] With early damage estimates for the basin reaching into the millions of dollars, residents in the lower basin called the June 1944 flood the "worst flood in history."[53]

At Stanton, 4.8 inches of rain caused a flash flood that crested at about 7 a.m. on Sunday. Meskenthine Creek, which flows into the Elkhorn near Stanton, apparently caused much of the local damage as it flooded basements, gardens, and croplands. Stanton County's lowland farmers felt the frustration of having their crops ruined twice in the same year—a problem throughout much of the basin in 1944. Even crops that had been harvested were not safe as floodwaters ruined stored grain. Livestock producers, too, suffered losses. Martin Melcher, who owned a farm on Union Creek, lost eighty-two hogs, three head of cattle, and some machinery. Roy Chamberlain on Butterfly Creek lost a pair of calves and one hundred spring chickens. Further losses came when the town, about midway on the washed-out rail route between Norfolk and Pilger, lost rail service for about a week. Washouts blocked highways, too.[54]

Pilger sustained minor damage on Sunday morning when Humbug Creek breached a dike south of Highway 275 and drainage channels overflowed from runoff produced by an estimated eleven to fourteen inches of rain. Most homes escaped serious water damage, but bottomland farmers lost crops. That evening, the Elkhorn overflowed at Stanton on its way to a record setting height.[55]

Further downstream at West Point, the storm dumped 7.2 inches of rain between 1 and 8 a.m. Sunday. Other portions of Cuming County reported up to ten inches. Between 7 and 7:30 a.m., West Point's fire whistle cut through the faltering rainstorm to warn residents of the impending flood. Runoff from the hills around West Point brought flooding to the town's western neighborhoods. At about 8 a.m., the Elkhorn edged above flood stage (ten feet), fueled in part by the flood-like runoff draining from town.

Plum Creek's excess waters arrived in town at approximately 10 a.m., flooding Neligh Park and the county fairgrounds, both in the northwest part of town.[56] Plum Creek also severed traffic into West Point on highways 275 and 32. Its flow added to an already swollen Elkhorn and by noon the river had climbed beyond twelve feet. By late morning on Monday the 12th, floodwaters at West Point nearly reached the high-water mark set in 1940. Although the river did not reach its old mark, the flood blocked all roads into town and continued rains in the vicinity left residents expecting a second crest late Monday or early Tuesday. At its crest on Monday afternoon, the Elkhorn reached 14.6 feet at West Point and measured more than a mile and a half wide. By Tuesday the receding river had dropped to thirteen feet.[57]

Elkhorn tributaries wrought heavy damage on a number of towns as heavy rains caused normally docile creeks practically to explode from their banks in flash flooding. In Colfax County, a flooded East Fork Creek hit Howells leaving fifty residents homeless and damaging property in at least forty homes. When the Middle Fork of Maple Creek flooded nearby Clarkson, it swept at least five houses off their foundations. The wall of water carried three of the houses three-quarters of a mile downstream. The houses of the John Sodomka and Ed Malena families proved seaworthy as they carried the families out of town, across the railroad tracks, and about a mile into the countryside before they settled into the muddy bottomlands where the stream widens. The devastation left twenty-four families homeless and inflicted property damage estimated in the many thousands of dollars.[58]

One of the flood's most bizarre reports came from nearby Howells, located on the East Fork of Maple Creek. Six members of the Jerry Spatz family fled to their home's second floor as water entered the first floor. In the excitement, the family did not notice that the current had pulled their home from its foundation and was carrying it, and them, downstream. As the Spatz residence headed for what appeared to be an oncoming building (which was, in fact, stationary), Mr. Spatz warned the rest of the family "Brace yourself for a shock—a building is coming this way and will hit us."[59]

On Sunday the 11th, Maple Creek, swollen to "crushing power," ripped four houses from their foundations in Clarkson and carried two of them—occupants still inside—about a mile down the creek's channel. Although a flood warning had alerted residents of the impending deluge, the "ravening wall of water" hit so quickly that people were unable to evacuate. Eyewitnesses living in the Maple Creek valley eight miles east of Fremont likened the wall of water to a "white cloud" descending upon them and their home. In addition to buildings, the torrent swept away chickens, cattle, and horses. Before the flood crest passed town, the creek also undermined the foundations of other homes and washed out a half-mile of railroad track.[60]

Scribner, which lies in the floodplains of both the Elkhorn River and Pebble Creek, incurred heavy damage, mostly from the creek. Pebble Creek overflowed at 7 a.m. on June 11. The creek's waters climbed for five hours, cresting at noon that day. Two to three feet of water covered the entire city, with some areas under as much as six feet. Shortly after Pebble Creek began to retreat into its banks, the Elkhorn overflowed. The flood hit Scribner so quickly that few people had time to evacuate. The business district, streets, utilities, and 237 homes suffered varying degrees of damage sending some families to seek food and shelter at the city hall. When the damages had been tallied, the total came to $378,300.[61]

At about 7:30 Sunday morning Hooper's fire alarm warned Hooperites that high water was on its way to town. Reports of Pebble Creek washing over Scribner and of a four-foot wall of water rolling down the highway and railroad tracks toward Hooper put officials on guard. By noon, most of the town's residents had found safety on high ground, where some were able to watch waves of floodwater the height of a grown human crash through town. Muddy waters rolled into basement windows and crept up the steps and into most businesses in the downtown district.[62] On Monday, additional rains in the area left local residents expecting the river to rise again.[63] Additional crests apparently did not materialize, however, as the *Fremont Guide and Tribune* reported Tuesday that all communities were digging out from the wreckage strewn about the valley by the rushing waters of the Elkhorn and its tributaries, some of which were so small and (previously) insignificant that "no one had ever taken the trouble to name [them]." One such stream reached depths of thirty to forty feet during the flood.[64]

Damage and Relief

Bell Creek climbed to record heights as it flooded Arlington in Washington County. Waters reached the town's sewage system and pumping plant—the first flood ever to do so. The Elkhorn, which, along with seven miles of land, separates Arlington from Fremont to the west, hampered transportation and communication between the communities and within and across the basin as it swelled to approximately four-and-a-half miles wide between the towns.

Valley and Waterloo, too, found themselves under water ranging from waist-deep to five feet. Many Waterloo residents flooded out of their homes on Sunday were unable to return until the following Friday when water still covered some low-lying roads. Even by then there were still residents depending on the Red Cross for supplies.[65]

As the flood receded, increasingly accurate damage reports became available. A Red Cross survey in Dodge County, for example, disclosed extensive damages. Although the Platte River also flows through Dodge County and the survey did not specify the Elkhorn and its tributaries as the sole sources of flood damage, reports indicate that the Elkhorn was responsible for most, if not all, damage in the area. Of Dodge County's 341,779 acres, 92,480, or slightly more than 27 percent, were flooded. Land holding about 25 percent of the county's corn crop fell within the flooded area. The flood destroyed seven homes and damaged 743 more. It also destroyed 121 barns and outbuildings and damaged another 354. The county lost 1,500 miles of fencing and twenty-five pieces of farm machinery, and suffered reparable damage to about a thousand more pieces of farm machinery. About 161 businesses, churches, and schools also sustained flood damage. Livestock also suffered heavily. Estimates placed Dodge's losses at fifteen work animals (horses and mules), 456 head of cattle, 1,800 hogs and pigs, 225 sheep, and 120,000 head of poultry of various kinds.[66]

Besides presenting a direct loss to the farmers who had raised them, the cost of disposing of their animal remains added to the flood's costs. Thousands of animal corpses swept away by the rising floodwaters were deposited randomly throughout the flood area as waters receded. By June 18 animal corpses had been removed from Hooper, but many still lay strewn about the countryside. A State Health Department official suggested burning, burying, or rendering animal remains. While burying or rendering the corpses would provide valuable nutrients to the soil or a salable product, burning would likely have added to environmental problems, especially if burners followed the *Fremont Guide and Tribune*'s advice to use petroleum distillate or crank case oil to facilitate cremation.[67]

Many species of fauna thrived in the post-flood conditions, but certain ones heaped yet another problem on corn growers. Moist conditions after the flood created a healthy environment for seed corn beetles, which feed on the hearts of seed corn kernels. About ten days after the flood hit, farmers began reporting the quarter-inch long beetles in great numbers. A farmer, who planted fifty acres of corn one Thursday evening, lost most or all of that field to the beetles by the following morning. Extension agents urged farmers to pack the soil firmly in their cornfields to discourage the burrowing pests, thus creating more work for the flood-stricken farmers and increasing their losses at the hands of the Elkhorn.[68]

Damaged transportation networks in the basin extended the consequences of the Elkhorn flooding to the region. Fremont, separated from Omaha by the Elkhorn, could rely for a time on only one road, Highway 77, to carry vehicular traffic out of the city. No trains entered or left on the morning of June 12.[69] By early July State Engineer Wardner G. Scott had tallied the expense of repairing roads and bridges in the flood area. On July 10, Scott announced a $425,000 repair and restoration program for which he hoped to gain $212,500 in federal funds.[70]

The June freshet disrupted Chicago and North Western Railroad service in the basin. Nearly every mile of track between Arlington, near the Elkhorn's mouth, and Norfolk, the basin's commercial hub, suffered some destruction. The flood did its worst damage to tracks near Scribner, Beemer, and Arlington. Officials estimated repairs to restore service would take two weeks.[71] Yet in September, the Scribner-Oakdale line still had not been restored as railroad officials waited on supplies and labor to complete the $230,000 project. Service was still out, too, on the road's Scribner-Humphrey line, which had not carried traffic since before the June 11 flood.[72]

The floods also added stress to human psyches. Robert Leisy, a sixty-one-year-old farmer whose land south of Wisner flooded in June, shot and killed himself on the 14th of that month. Sources do not allow us to connect the floods directly to Leisy's suicide, but it's easy to imagine tribulations from the flood exacerbating other problems in his life. The flood damage might have provided the proverbial "last straw" that led him to take his life.[73]

The destruction wrought by 1944's record-breaking flood elicited the most comprehensive and concerted responses from residents, government agencies, and charitable organizations yet. Among the first relief forces to arrive on the Elkhorn flood scene was the Nebraska National Guard. Fremont's Company D rushed into the field shortly after the flood hit on Sunday. By Monday, two additional companies, E and F from Omaha, had joined them. The forces remained on duty for much of the next week. On Saturday the 17th, companies D and F were mustered out, but Company E remained on duty along with detachments of administrative, medical, service, and transportation personnel.[74]

Within ten days of the May flood, the Norfolk Drainage Board, acting on orders from the Corps of Engineers, began repairing the dike that ruptured on May 12.[75] While the June flood wrought damage on downstream towns, the *Norfolk Daily News* reported that the Drainage Board had not had a major project in two or three years. Fearing more flooding, Norfolk's residents worried that trees in the North Fork's channel would clog it and hamper drainage. The newspaper report suggested that the Drainage Board should remove the trees from the channel.[76]

The Red Cross played a large role in helping flood-stricken areas by supplying emergency relief and long-term rehabilitation. From its area headquarters in Norfolk, the Red Cross fed and housed families flooded from their homes. The agency also provided clothing and other material relief to help flood victims reestablish themselves. By May 22, 130 families had registered to receive aid from the Red Cross.[77]

Although many Red Cross volunteers were local residents, several supervisors and caseworkers came from the agency's St. Louis office. Money, too, flowed in from the national office's disaster relief fund. Much of that money went to projects such as repairing flood-damaged buildings and buying seed corn, livestock feed, fencing materials and other items necessary to get flood-stricken farmers back in business.[78] Red Cross assistance indicated that the cost of living with the Elkhorn, at least in terms of unnatural disasters related to living and working in the floodplain, were being paid with assistance from outside the basin.

While still completing relief work from the May flood, the Red Cross aided victims of the June flood. At its Fremont headquarters, the Red Cross daily prepared 4,500 meals, which it shipped to Scribner, Hooper, and other areas. More than a week after the flood struck, the Red Cross was still serving meals to between 100 and 150 people. The Hooper chapter set up a local first-aid station in a room at the Hooper Inn where all residents were to be inoculated against typhoid, which threatened areas with flooded wells. Nebraska's Health Department, assisting in relief, provided 1,200 doses of typhoid vaccine to Howells alone. The department also sent a water chlorination unit to provide potable water for the town.[79]

Other civic groups joined the relief effort. In Hooper, for example, the board of trustees was joined by Red Cross workers, volunteers from the American Legion and Auxiliary, the fire department, and members of the Commercial Club. But membership in a civic group bore no significance as volunteers of all stripes came out to help their neighbors get through the disaster. One recovery technique that appeared unique to Hooper was a village "lost and found" set up on an empty lot. People were encouraged to bring items that had washed onto their property to the lot where, authorities hoped, the owners would claim them.[80]

Norfolk radio station WJAG provided stopgap communications relief in coordination with Northwestern Bell Telephone while phone lines were out of order. Beginning at 8 a.m. Sunday the station broadcast flood bulletins and messages, an important service to areas where phone service had been cut by the flood. For example, the station requested boats be sent to Howells.[81]

One response indicated that floodplain development had not been well-planned. The roadbed of the highway—likely US 77, which crosses the Elkhorn north of town—into Winslow dammed floodwaters for miles and exacerbated flooding in that town. An eight-foot–wide hole was blown in the highway to allow the water to drain. A side effect of that treatment, obviously, was the interruption of traffic. Ironically, the Elkhorn's destructive power *saved* Nickerson from severe flooding when high water washed out a section of railroad grades that had also acted as a dam and caused Maple Creek to back up. When the grade gave way, dammed waters rushed through the gap and into the Elkhorn thereby minimizing the potential community damage.[82]

In at least one place, Snyder, located along one of the unnamed creeks in Dodge County, the response was no response—at least from outside agencies. Left without telephone service or electricity for two to three days, Snyder's residents did for themselves, boiling their water and assisting the needy with food, clothing, and shelter. The Red Cross eventually arrived to provide typhoid vaccinations.[83]

A New Era of Intervention

In response to the tremendous damage caused by the 1944 flood, the local, state, and even national government responded like never before. They supplied immediate relief to the victims and sought to repair the damaged roads, bridges, and buildings. At the county level, for instance, Cuming County's Board of Supervisors, as it had in 1940, juggled its budget to meet the unforeseen costs of repairing roads and bridges. Supervisors increased the Bridge Fund from $4,412.74 to $35,921 and more than doubled the amount in the Road Fund, from $15,694.06 to $33,766.63. As in 1940, supervisors raised the money by increasing taxes, spreading the burden for 1944's food damages throughout the county.[84]

The state came to the aid of flooded famers by helping them plan replanting strategies and to obtain seed. D. L. Gross, an agronomist from the University of Nebraska's College of Agriculture, visited several flood-stricken farms to issue advice. In Dodge County he told corn growers that if corn was up and not covered with mud, it would survive if floodwaters were not too slow in draining away. Growers were probably disappointed to learn that Gross could not tell them how slow "too slow" was. On the other hand, they were probably relieved to learn that Gross's office raced against the clock to canvass the state for all available seed for replanting. Gross estimated that if growers could replant their fields with early maturing hybrids by June 20, their crops would likely have a chance to mature. The extent of flood damage to fields prevented planting any other crop, and Gross recommended that as many growers as possible stay with corn for the season. Where this was impossible, he advised leaving flooded areas unplanted during the summer and planting wheat in the fall.[85]

The federal government responded at several levels. The US Weather Bureau, for instance, established a flood warning system.[86] More important, the federal government, led by the Corps of Engineers, coordinated a number of local, state, and federal agencies, in cooperation with residents, to research and analyze flooding and flood damage in the Elkhorn Basin. The results of this inspection reflected recent changes in bureaucratic perspectives on river management and would change the government's stereotype of the region and focus federal resources on flood control.

The severity of 1944's flood damage challenged the US government to reevaluate its assessment of the Elkhorn Basin as part of the generally flood-free Platte River Basin. Two documents are especially important to federal involvement on the Elkhorn. First, a 1934 document, known as the 308 Report on the Platte River—one of about two hundred such reports on US streams prepared as called for in the Rivers and Harbors Act of 1927—contained the government's position on the Elkhorn. The 308 Report was a comprehensive review of the Platte Basin, which includes the Elkhorn and its tributaries, examining irrigation, water power, and flood control, which concluded that the Platte River Basin was not subject to destructive flooding.[87] Second, the Flood Control Act of 1936 established federal government responsibility for flood control on rivers and their tributaries, which, when flooded, caused loss of life and property, eroded land, and impaired or obstructed navigation, highways, railroads, and other channels of interstate commerce, thereby constituting a menace to national welfare.[88] Taken together, these documents reflect the workings of the federal government expanding its reach and vision with respect to natural resources. They reveal a mindset tuned to the utility of natural resources and to viewing flooding as a disaster, a menace to the state. And they reveal a worldview calibrated to using federal resources to marshal technological know-how to manipulate rivers to protect a national welfare that by virtue of its organization depended on interaction with rivers in their floodplains. The Elkhorn River Basin or other Plains river basins were not the only places where people had staked their lives on real estate that wasn't dependably dry. Furthermore, the 308 Report poses a standard, "destructive flooding," that would seem ripe for debate. By 1944 the Elkhorn had for decades produced what basin residents would certainly characterize as destruction. The floods of 1944 re-opened conversations on these matters.

A political climate favorable to reviewing the Elkhorn's status also played a vital role in the change. The summer of 1944 found Congress embroiled in the Missouri Basin Project, better known as the Pick-Sloan Plan, which called for reclamation and flood control projects throughout the Missouri Basin, which included the Elkhorn. Following severe flooding on the Missouri River in 1943, the House Flood Control Committee asked the Corps to review the Missouri River for flood control options, a task the Corps assigned to Col. Lewis A. Pick, then division engineer of the Missouri River office of the Corps of Engineers, stationed at Omaha. In about ninety days, Pick composed a response that employed multiple-purpose dams to control flooding and provide possible irrigation and power production. This marked another step in the expansion of the Corps' duties. Until 1917 the federal government supported levee construction only insofar as it benefitted navigation, an approach which changed that year to include support for levees meant solely for flood control. Eventually the Corps got into the business of flood control on the Mississippi and, in the 1940s, on the Missouri. W. G. Sloan, an engineer in the Department of Interior's Bureau of Reclamation, had also been working on a development plan for the Missouri. Congress received both plans, Pick's emphasizing flood control, Sloan's emphasizing irrigation and hydroelectric power production, within months of each other in February and May 1944, respectively. As the Elkhorn flooded in 1944, Congress was in the midst of studying the two plans, which it would eventually fuse.[89] The time was ripe for flood studies, especially those concerning the Missouri Basin.

On June 21, 1944, the House Committee on Flood Control ordered a survey of the Elkhorn Basin to determine if federally financed flood-control measures were justified.[90] Beginning with a pair of flood damage hearings, on August 29 at Norfolk and on August 30 at Fremont, the Corps would review the conclusions of the 308 Report, evaluating the Elkhorn's flooding through firsthand consultation with local residents.[91]

In preparation for the hearings, several towns held organizational meetings to discuss the flood's damages and to explore options for preventing or limiting future damages by the Elkhorn and its tributaries. Hooper held one of the earliest such meetings on June 23 when forty to fifty people representing the county extension department, the Agricultural Adjustment Administration, bankers, newspapers, the Red Cross, the Farm Security Administration, and the Reconstruction Finance Corporation. It included mayors or other city officials of every Dodge County town except North Bend who met to discuss flood issues.[92] With this kind of preparation, the Corps was certain to get an earful.

On Tuesday, August 29, about 150 representatives of towns, civic groups, and farmers from Pierce, Cuming, Stanton, Antelope, and Madison counties related their experiences at the Norfolk hearing. According to testimony, 1944 flood damages came to more than double the damages suffered in the 1940 flood. Losses were especially heavy, in part, because the 1944 floods happened so late in the growing season that some farmland had to go unused, whereas in most previous floods, land was successfully replanted after waters receded. Russell Young, a farmer from Tilden with thirty-four acres of bottomland, testified that it "produces real good every year," but that "we never had anything this year at all." Young went on to tell the Corps that the floods occurred often and "every one seems to be getting a little worse."[93]

B. E. Adkins, a banker from Norfolk, addressed the hearing on behalf of fifty to sixty Madison County farmers who worked land along the Elkhorn. Prior to the hearing he met with the farmers and asked them to fill out a questionnaire regarding their flood experiences going back as many as sixty years. Although he had not collected all the questionnaires by the time of the hearing, he estimated that the group represented 6,450 acres, which sustained $81,900 damage in the 1944 flood. A consensus of the farmers agreed with Russell Young that floods had been getting progressively worse during their years on the Elkhorn.[94]

Overall, those testifying believed that the Elkhorn and its tributaries should be straightened and widened and that reservoirs should be built at tributaries' headwaters to slow the flow of water into the main stem during flood situations. Generally speaking, laymen, mostly farmers, advocated local channel modifications and levees while those with technical training advocated similar responses on a basinwide scale. "Elkhorn—that's the right name for it. . . . My suggestion would be to take some of these worst bends out," E. C. Diers, a native and longtime farmer and cattleman from Fremont, suggested.[95] Wardner G. Scott, state engineer, after reporting his estimate of damages to highways and bridges at the hearing, recommended a basinwide plan for flood control to protect the transportation infrastructure. M. J. Hankins, district conservationist at the Stanton Soil Conservation District in Stanton, also recommended a basinwide approach, although of a different sort. Hankins pointed out that while floods hit only about 30,000 acres of Stanton County's total area of 276,000 acres, runoff from the unflooded 246,000 acres created the flood. To minimize future flooding, he recommended soil conservation measures to hold rain water on the land where it fell.[96]

The next day in Fremont, testimony from lower-basin residents echoed that of their upstream neighbors. They recommended that the Corps straighten and widen the Elkhorn's channel and build dikes and levees to protect low areas. The crowd realized, however, that it would be virtually impossible to prevent a disaster similar to that which resulted from the intense flash rains of June 1944. In some cases, they acknowledged, nature just could not be contained.[97] It could, perhaps, be managed so that floods would not be as devastating.

Days after the Norfolk and Fremont hearings had been completed, the *Fremont Guide and Tribune* editorialized that localized control projects were not the answer. Local protection projects might successfully save one town from the ravages of a flood, but that town's dikes and levees would merely divert the water to another location which, if unprotected, would suffer more than it would have otherwise. What the basin needed, according to the *Guide and Tribune*, was a comprehensive, basinwide management plan.[98]

Engineering and the Elkhorn

The government apparently listened to the advice given at the Norfolk and Fremont hearings. In the following years, the Corps of Engineers surveyed the entire basin conducting an inspection of existing protection projects and the sites of all specific improvements suggested in testimony at the hearings.[99]

But even before engineers began their survey, Congressman Karl Stefan of Norfolk told his constituents that the federal government did not have funds available for extensive flood control on the Elkhorn. Stefan promised to visit the flooded locales with an engineer from the Omaha district of the USACE and that an official Army survey would be made. From the survey, Stefan told members of the West Point Community Club, individual towns would be able to draw suggestions for local flood protection projects. But Congress, he told them, had little money available except to finance minor construction work such as "a small dike here and a bit of revetment there."[100] Such structures would raise, but essentially preserve and stabilize the existing banks, while allowing the river's channel to contain greater volumes of water. Stefan thought in terms similar to those of the Elkhorn's farmers and ranchers when it came to how the basin might be shaped. Furthermore, given that World War II still raged, 1944 was not a good time to seek Congressional funding for a problem that did not pose an existential threat to the nation.

Portions of the basin, including the upper reaches of the Elkhorn, the upper reaches of the North Fork, and the upper portions of Cedar, Logan, Union, Humbug, Plum, Pebble, Maple, Rawhide, and Bell Creeks, merited no flood control measures in the eyes of the Corps. Some of these areas simply did not flood, or did not flood often enough or seriously enough to warrant expensive modifications. In other areas, so-called serious flood damages occurred only during the 1944 floods. In some of these places, notably in areas whose elevations placed them above possible flood damage from streams, damages resulted from the intense nature of the rainstorms and from overland runoff rather than from streams overflowing their banks.[101]

Using information gathered from the hearings and surveys, the Corps drafted four comprehensive plans of improvement and a number of local protection projects that could be undertaken individually. A brief description of each plan provides insight to the state of the art and the art of the practical when it came to mid–twentieth-century flood control.

Comprehensive Plan A called for improvements to the Elkhorn that would protect against floods of the magnitude of a "super flood" comparable to the 1944 inundation from Neligh to the river's mouth. On tributaries, Plan A called for a "high degree of protection," a level that would protect against floods similar to the floods of record on the tributaries. Levees and channel rectifications would provide the means for protection along both the Elkhorn and its tributaries.[102]

Plan B aimed at flood protection along the Elkhorn from Norfolk to the river's mouth against floods about half the magnitude of the record 1944 flood and to provide "a high degree of flood protection" on tributaries including the North Fork from Pierce to its mouth and the lower reaches of Giles, Buffalo, Battle, Meskenthine, and Pebble creeks through the use of levees and channel "rectification." Along the main stem of the Elkhorn, the plan called for a floodway with levees and "channel rectification to obtain a reasonably smooth alignment by elimination of large and objectionable bends." Essentially the same as Plan A, Plan B would merely provide protection against floods of lesser magnitude and would not include the area from Neligh to Norfolk.[103]

Plan C like Plan A, provided flood protection from Neligh to the Elkhorn's mouth and along tributaries against floods of the magnitude experienced in 1944. Plan C, however, proposed a flood control reservoir on the Elkhorn below Scribner to supplement levees and channel rectifications in other areas. Plan C's reservoir would hold 330,000 acre-feet of water and would have a surface area of about 16,400 acres, which would require relocation of the towns of Scribner and Crowell, 23.5 miles of railroad track, and 24.75 miles of Highway 275.[104]

Nearly identical to Plan C, Plan D called for a larger, multi-purpose reservoir at Scribner that would allow for not only flood control, but also for hydroelectric power generation and irrigation. Plan D's reservoir would hold 800,000 acre-feet of water covering about 26,000 acres and would require the same towns and roads to be relocated as the Plan C reservoir. Of its capacity, the Plan D reservoir would utilize 330,000 acre-feet for flood control and the remainder for irrigation and hydroelectric power generation.[105]

The Corps also suggested a series of twelve local flood protection projects that could be undertaken by cities or towns in partnership with the federal government. These projects, which were also components of the comprehensive plans, would allow residents to protect strategic portions of the floodplain, usually towns, and avoid the extraordinary expense of improving the entire length of the river. Areas recommended for projects included the North Fork at Pierce and Norfolk; Giles, Buffalo, Battle, and Meskenthine creeks; and the Elkhorn at West Point, Scribner, Hooper, Winslow, Waterloo, and in an area above the river's mouth.[106] Some projects involved levees and channel diversions or modifications. For example the town of Pierce would be protected against floods equal to the magnitude of the town's worst flood by a two-mile long levee on the south, east, and north sides of town. In addition, about 1,700 feet of the North Fork's channel would be enlarged and relocated south of the town's park to handle flows that would normally overwhelm the capacity of the Pierce Dam. The dam's spillway and downstream channel would also be modified to improve their drainage abilities.[107]

Likewise, the flood protection system recommended for Norfolk and the agricultural area east of the city included levees and channel diversions, which would keep the city dry during floods as bad as 1944's record one. Complications arose from existing improvements along the channel, which left insufficient space for levees, from the existing power dam and its water rights, and the high probability that if a flood occurred that exceeded the capacity of the proposed modifications there would be immense damage to objects supposedly protected by the works. Also, the walls needed to contain floods would be too large and too expensive to build. To address these problems, engineers adopted the suggestion of residents to create a diversion channel for the North Fork.[108]

The Corps suggested lining the existing North Fork channel with levees and lining the diversion channel to be built east of the city. North of town, in the natural channel, it recommended a control structure with a gate that let humans direct some water into the existing North Fork channel, but also send excess flows into the diversion channel away from homes and businesses in town.

From the control structure, the diversion channel would run southeast for about 6,000 feet, turn south for about 4,700 feet, and rejoin the original channel. Spring Creek, too, would be diverted to join the North Fork diversion channel about 2,000 feet above its mouth. Where the natural and diversion channels met, the North Fork's channel would be widened and a levee-enclosed flood channel built around it. Farms, or portions of farms, within the floodway would be moved. The banks of the original channel, where it flowed through town, would also be secured with levees. Finally, to protect south Norfolk from the Elkhorn, the plan called for a levee along the river. The project would require abandoning a number of old roads, building new roads, and raising some railroad grades.[109]

Most of the local protection projects, however, consisted simply of minimal channel modifications and levees to be built around cities as bulwarks against high water. Such would be the case for agricultural areas along the North Fork from Pierce to the mouth. The North Fork's channel would be widened in places and straightened where "objectionable" bends constrained flow.[110]

The Corps proposed a more aggressive project on Giles Creek, from Tilden to the mouth, to protect Tilden and riparian agricultural lands from floods equal to the worst recorded flood on the creek. The Corps suggested blocking the original Giles Creek channel west of Tilden and sending the creek's flow into a levee-enclosed diversion channel about 9,700 feet long, which would empty into the Elkhorn. An especially steep gradient through the town of Stanton earned Meskenthine Creek, too, a recommendation for aggressive treatment from the Corps. Since the channel's steep fall creates the potential for flood flows of destructive velocity, the Corps proposed to line the channel with concrete where it flows through the city—about 3,700 feet from the north side of town to the south. The concrete-lined portion of the channel would also be surrounded by six-foot high levees. Along the creek's remaining run to the Elkhorn, another 2,000 feet, the Corps suggested enlarging the channel.[111]

To protect Meadow Grove from Buffalo Creek flooding, the Corps suggested a 5,600-foot-long, four-foot high levee and the elimination, through making a cut-off, of an "objectionable bend." Battle Creek, too, would be straightened. Nine cut-offs would be required to make the stream's channel unobjectionable. About 10,800 feet of levees built on its right bank and another 5,600 feet of levees built on its left bank would protect the town of Battle Creek.

Pebble Creek, Scribner's flood nemesis, earned the most purely defensive recommendation from the Corps—a system of levees from 1.5 miles west of town to the Elkhorn's mouth. Nearly a mile of existing levees would be rebuilt to offer better flood protection to the town and adjacent agricultural areas. Projects at West Point, Hooper, Winslow, Waterloo, and above the Elkhorn's mouth would also rely primarily or entirely on levees for protection.[112]

While straightening and enlarging the channels of the Elkhorn and its tributaries attracted wide support among residents, engineers found that implementing such improvements basinwide would be economically infeasible. In making that decision, the Corps relied upon estimates of past and future flood damages on the Elkhorn and its tributaries. It should be noted that estimates for future damages often exceed estimates of past annual damages by multiples. The Corps predicted an increase in the severity of flood damage through an increase in the frequency and magnitude of flooding, an increase in floodplain development and the attendant higher potential for damage, or a combination thereof. In any case, the figures showed that much of the Elkhorn's flood-prone segment—from Neligh to the mouth (basically the lower basin)—was rural, making basinwide improvement inefficient (or cost ineffective).

Why spend money to control the river where people and valuable assets were not in harm's way? Furthermore, the Corps speculated that flood-proofing such regions would not encourage enough development of those areas to merit the expense of protecting them. Although the rural areas held rich agricultural land and would incur high future flood damages, the Corps deemed flood damages in those stretches too diffuse to justify the cost of basinwide flood control measures. But near towns and cities, flood control measures would provide the most favorable cost-benefit ratio.[113] Here it might be said the nature of life on the Great Plains came home to roost. It made engineering and ecological sense to treat the basin as a unit, but it did not make economic sense to spend money on the least-densely populated portions of fly-over country. At the same time, the recommendation to protect the region's most densely populated and richly developed places showed how decision-making for the Great Plains echoed river development and flood control measures for densely populated urban areas such as New Orleans and Los Angeles.

The Corps also determined that flood control via reservoirs on tributaries as suggested at the Norfolk and Fremont hearings was infeasible because few suitable sites for dams and reservoirs existed and because the drainage areas and reservoirs at suitable sites would be small in proportion to the size and cost of the dams required. If those reasons were not compelling enough, the Corps also found that tributary reservoirs would have no appreciable flood control effect on Elkhorn River floods.[114] Engineers did, however, include minor channel modifications and levee construction—where feasible on a local basis—in their report.[115]

Addressing the proposed plans purely in terms of flood control, the comprehensive plans seem to offer the best approach to minimizing flood damages along the Elkhorn. The logistics of drainage, runoff, and flooding within a river basin system would indicate that a piecemeal approach would fail or create problems in some areas as they are solved in others. The history of the Elkhorn's drainage districts confirms that. Successful reduction of flood damages along Logan Creek came from modifying nearly the entire length of the creek and maintaining the improvements over time. In areas where only portions of streams were modified or where modifications were allowed to deteriorate, such as those along Rawhide Creek, results were less successful. Also, the demise of some drainage districts in a quagmire of litigation showed that saving one location from flooding often meant harming another with increased runoff.

Furthermore, flood studies show that a stream's floodplain should be treated as a complete unit. Engineers now operate under the assumption that building within the floodplain increases flooding. Visualizing the floodplain as a paper cup aids in understanding this concept. Under normal conditions the river flows within its banks, or the analogous paper cup is about half full of water. When the river floods, it fills its floodplain going to the top of its natural channel, or the paper cup fills to the rim. But building in the floodplain displaces the river's normal flow forcing floodwaters out of the floodplain. Using the paper cup analogy, building in the floodplain has the same effect as if a person holding the cup squeezed it, thus forcing excess water to spill over the sides.[116] That analogy illustrates the virtue of comprehensively isolating a river's floodplain—the true limits of the stream's channel as discussed in Chapter 1—for purposes of flood control.

Environmentally speaking, treating the entire floodplain as a unit also makes sense. As the preceding discussion shows, encroachment into a stream's floodplain forces flood flows higher than they would normally reach. Thus humans not only subjected their structures in the floodplain to water damage, but also extended damage to structures that otherwise would have been beyond the natural floodplain. Aside from vacating the floodplain, which would appear to be out of the question given the amount of time and money invested in it, limiting human intrusion and giving the river its berth seems a reasonable course. These ideas were ahead of their time.

But politics and economics complicated the world into which the Corps' recommendations came. Humans had already intruded into the floodplain, and for nearly a century had become ever more firmly entrenched and committed to remaining on the floodplain. To implement the comprehensive plans would require cooperation from myriad local governments to gain rights-of-way and, in the cases of plans C and D, uprooting two towns completely. Implementation would also require allocating exorbitant sums of money for comparatively insignificant returns. The total cost of constructing Plan A would have been $34,366,500; Plan B, $29,198,800; Plan C, $76,138,900; and Plan D, $84,478,200.[117] Once constructed, the economic benefits of these projects would be exceeded by their maintenance costs (see Table 4-3).

Table 4-3: Costs and Benefits of Comprehensive Plans for Flood Control on the Elkhorn River

Plan	Cost	Annual charges	Annual benefits	Cost-benefit ratio
A	$34,366,500	$1,541,066	$710,660	1:0.46
B	$29,198,800	$1,300,726	$561,450	1:0.43
C	$76,138,900	$3,327,912	$697,760	1:0.21
D	$84,478,200	$3,740,946	$923,310	1:0.25

Source: U.S. Army Corps of Engineers, Omaha District, *Review Report on Improvement for Flood Control and Allied Purposes, Elkhorn River and Its Tributaries, Nebraska* (Omaha: Department of War, 1947), 99–100.

Given the complications associated with the comprehensive plans, local protection projects became the most economically and politically attractive avenue to minimize flood damage. These projects were comparatively inexpensive and, in most cases, provided estimated annual flood control benefits greater than their annual maintenance costs. Most important, such projects concentrated on small areas and protected concentrations of wealth and people—municipalities.[118]

In some cases, the local protection approach made local cooperation easy to obtain when the Corps presented its research and recommendations to basin residents in hearings held at Fremont and Norfolk in November 1946 and in subsequent meetings with local officials in the areas affected by the local projects. Various city and county councils endorsed most of the local plans, which meant committing to provide land, rights-of-way, and easements at no cost to the federal government, releasing the federal government from liability stemming from damages due to the construction work, agreeing to make necessary alterations to highways and bridges affected by the projects, and agreeing to maintain the projects after their completion.[119]

Yet in other cases, the decision by local governments not to participate in the recommended project indicates that more than economics was involved. For example, the projects recommended for Buffalo Creek (1:3.89), Scribner (1:2.09), Hooper (1:1.38), and Winslow (1:1.15) offered some of the most favorable cost-benefit ratios available. The village board of Meadow Grove rejected the Buffalo Creek project because it could not obtain waivers from landowners along the creek who believed they might be injured if the project were completed. The reluctance of Meadow Grove's citizens to agree to the plan might have come from their opinion that their flood troubles stemmed from highway and railroad embankments interfering with Buffalo Creek's drainage. At the Norfolk flood damage hearing, Meadow Grove's mayor, Leonard Sanne, asked for a ditch to be cut through the embankments to allow water to drain on its natural course. With an apparently simple solution at hand, people in Meadow Grove perhaps believed that the Corps proposal was too elaborate. Scribner's council simply believed its project was undesirable. Hooper's village board rejected the project slated for its jurisdiction because it feared high maintenance costs and legal tangles that might result from the project's construction. Winslow sent the issue to its citizens, who rejected it because of its expense and the feeling that the project was unnecessary.[120]

Conclusion

Environmentally speaking, curbing flood damage in a piecemeal fashion through local protection projects seems less desirable than a comprehensive approach. It treats the floodplain neither as a unit nor a natural part of the river. The inconsistency of this approach protects some areas of greater human wealth, yet leaves open to further injury areas of lesser human wealth. Further, and for the same reasons, such an inconsistent approach does not bode well for the natural mechanisms of the river and riparian lands and systems as water is artificially channeled in some places and not in others. Norfolk's city engineer, H. H. Tracy, put it nicely when he testified at Norfolk that "in channel straightening there is no place in any one locality to commence work or to stop, as the entire river is a series of convolutions across county after county."[121]

These solutions call to mind the environmentally unsound, shortsighted use of technology condemned by historian Donald Worster. In his book *Dust Bowl: The Southern Plains in the 1930s*, Worster criticizes inappropriately applied agricultural technology for exacerbating the Plains' natural drought cycle during the 1930s to create the Dust Bowl. The "ecological insensitivity of our culture," as Worster puts it, was also at work in the development of the Elkhorn Basin. To make matters worse, cultural insensitivity to the environment, at work from the beginning, may have been heightened by involvement of the federal bureaucracies, which Worster describes as "innately antiecological" because they are "too insulated from the results of their actions to learn, to adjust, to harmonize."[122] Rather than treat the river basin as a natural system that includes events known as floods, the twentieth-century approach to the river measured costs and benefits according to an unnatural landscape humans had imposed on the Elkhorn Basin starting in the 1800s.

Of the twelve local protection projects recommended by the Corps, only six, those at Norfolk, Pierce, Giles Creek, Battle Creek, West Point, and Waterloo met the dual criteria of cost-effectiveness and local support that would lead the Corps to recommend to Congress that it assist residents in curbing floods and flood damages in the Elkhorn Basin, which the Corps did in submitting its findings to the House of Representatives.[123] This result reveals decision-making and policy-making at work on the Great Plains where an increasingly complex mix of local, state, and federal governments had to figure out what to do and how to do it with respect to flood control even as the dominant ethos of living with the Elkhorn continued to be dominated by the assumptions that people, buildings, and infrastructure would remain in the floodplain and human will could be imposed, in some fashion to be determined, on the river.

The most comprehensive, thoughtful, and intensively researched response to flooding in the basin's history, then, did not attempt to dominate the river. Rather, it limited action to defend areas of concentrated capital where local interests would support the response. In light of the benefits of a comprehensive and coordinated flood control plan, the hit-and-miss democratic application of the Corps' cost-benefit–driven recommendations makes the response seem almost haphazard. Responses applied according to economic and political expedience limited manipulation of riparian lands, but failed to treat the floodplain as a whole. In the short run, that may have been the best result for the river system and may have inadvertently prevented further bureaucratic ecological insensitivities of the type Worster criticizes. In the long run, it left the floodplain open for uncoordinated development with disregard for the floodplain system and vulnerable to the cultural insensitivies about which Worster writes. It would be years before government agencies would adopt a comprehensive approach to floodplain planning.[124] But change was on the horizon. So were more floods.

Chapter 5

YEARLY FEARS OF FLOODING: LIVING WITH THE ELKHORN SINCE 1944

"It's a mean river."

Bud Lind, Elkhorn Basin resident[1]

The decades following the devastating floods of 1944 brought partial implementation of the flood control projects carefully planned by the USACE. Of the six local protection projects recommended by the Corps, all or portions of five were eventually built. The Pierce and West Point projects were started in 1963 and completed in 1964. Construction on the Waterloo project began in 1966 and was finished late in 1967. Work on the Norfolk project commenced in May 1966 and was completed in December 1968. The levees proposed for the Battle Creek project were abandoned in favor of 800 feet of riprap (large chunks of rock or similar material stacked on an embankment to prevent erosion), completed in March 1973, along both banks of the creek to protect the embankments for State Route 121's bridge. The federal government abandoned the Giles Creek project in November 1977.[2]

All or portions of two other local protection projects designed by the Corps, but not recommended for construction, were later built. A levee surrounded Hooper as of November 1966. Buffalo Creek at Meadow Grove sported an improved channel for spring of 1973.[3] Other Corps projects in the basin include a stone embankment to protect a bridge abutment on the Elkhorn River near Norfolk (completed in May 1973), channel enlargements to Union and Taylor creeks near Madison (completed May 1966), levees along Middle Fork Maple Creek near Clarkson (completion date not available), a bank stabilization project on Logan Creek near Bancroft (completed in February 1972), and a 120-foot riprap on the Elkhorn's right bank to protect the Q Street bridge abutment near Gretna (completed in February 1972).[4]

Despite all that work, which added to generations of projects to straighten, deepen, contain, and otherwise control the Elkhorn, the contest over the floodplain remained a constant concern for basin residents into the twenty-first century. Settlers who had staked claims to the floodplain, the borderlands between wet and dry, left a legacy of anxiety and regular damage to their heirs. A 2010 newspaper story about the spring river conditions led with this sentence: "Impending rain and melting snow have raised yearly fears of flooding in much of eastern Nebraska." After a century and a half of settlement and "improvement" of the landscape, the Plainsfolk of the Elkhorn Basin lived under a regular—annual, if the newspaper report had it right—fear of flooding.[5] Dealing with that fear and the reality of high water brought more attempts to bend the river to human will, application of resources from new agencies, and new approaches to living with the river that showed people might finally be getting wise to the Elkhorn's ways. Through it all, the floods kept coming.

The Floods of 1947

The floods of 1944 remained fresh in the minds of sons and daughters of the Elkhorn as they managed their relationship with the river in the spring of 1947, even as the federal government prepared its elaborate plans. At Norfolk, the drainage board laid plans to speed the North Fork's flow with the hope of avoiding damage on the scale witnessed three years earlier. Large trees would be removed from the North Fork's banks from the East Norfolk Avenue bridge (near downtown) to the stream's mouth. Although trees might be considered "natural" residents of the floodplain, they slowed the flow of high water. They had to go. The board also planned to deepen the river's channel, also to speed flow, by cutting through sand bars. Finally, the board sought to stabilize banks near vital bridges with riprap.[6] Whatever progress came on these projects, intense spring rainstorms made that spring and summer another season of flooding in the basin.

The 1947 floods didn't bring massive destruction, but they disrupted transportation and hurt prospects for lowland farmers. Torrential rains lashed the lower basin in mid-May, causing flash flooding and a nearly two-mile washout of the Chicago and Northwestern tracks just west of Dodge.[7] Another storm in early June caused flash flooding that derailed an eight-car train near Nickerson and washed out Chicago and Northwestern tracks in several places in the lower basin. At Norfolk, the Elkhorn flowed into yards in the southern part of town.[8]

Intense storms soaked the basin again in early June dumping as much as five inches of rain in less than three hours in some places and causing flash flooding that damaged property and transportation infrastructure. At Battle Creek, where some of the heaviest rain fell, more than half the town was under water. Basements flooded and some basement walls crumbled. High water ruined lowland corn planting and sent farmers back to fields to replant. Flooding washed out Chicago and Northwestern tracks near Nickerson and Snyder. The biggest disruption hit the railroad's line connecting Fremont and Norfolk when an undermined roadbed gave way and derailed two baggage cars, two coaches, and a Pullman car at about 6:45 a.m. on Sunday, June 1. Some passengers required hospitalization for injuries sustained in the derailment. Others found themselves welcomed into nearby farmhouses for breakfast before being returned to Fremont by bus. Near Snyder, Pebble Creek rose and washed out the approach to the Highway 91 bridge over the creek.[9]

The early June flooding coincided with the announcement of new measures to cope with high water. The Payne Creek Drainage District awarded a contract to "improve" the creek, which empties into the Elkhorn downstream from Stanton, by building a drainage ditch with dikes on each side. This aimed to keep the creek's floodwaters off several hundred acres of cropland, which were often damaged by heavy rains, and to get that runoff "safely" into the Elkhorn. The Chicago and Northwestern simultaneously raised its roadbed in the area and installed a new steel culvert in an effort to keep floods off its tracks. Farmers in the Stanton Soil Conservation District joined the flood fight through terracing and other erosion control measures designed to control rain where it fell, or slowing its journey to the creek.[10] Those concerned hoped to protect crops, improve transportation, and enhance the value of real estate by managing rainfall and improving drainage to minimize flooding.

Flash floods hit yet again in the middle of June thanks to a storm that dumped four inches of rain in parts of the basin. For the fourth time in two weeks, high water washed out the Chicago and Northwestern tracks near Dodge.[11] The Elkhorn ran over its banks for a ten-mile stretch near Beemer during the second week in June. The swollen river cut a new channel, which threatened to make Beemer's Elkhorn bridge obsolete. Upriver, eight miles southeast of Norfolk, the Elkhorn swept a wooden bridge into its current and slammed it against the Ertzner Bridge, a 300-foot steel span. Pressure from the swollen river pushed the debris against the Ertzner with enough force to tear the north end from its moorings and dump it into the Elkhorn, cutting traffic across the river at that point.[12] In the upper basin, the Elkhorn flooded lowland farms and ranches stranding livestock and leaving several farmsteads as islands amidst the high water. The relatively late-season damage to cornfields left some farmers unwilling to replant flooded bottomland.[13] Southwest of Stanton, Union Creek damaged 2,000 to 3,000 acres of farmland, leaving people in the area "anxious that something be done to eliminate the immense amount of damage done by the floods which have occurred so often in recent years."[14]

About a week later yet another massive storm dumped between five and seven inches of rain on the upper basin in a matter of hours setting off flash floods from Inman to Neligh. The floods washed out railroad bridges and those carrying highways 275 and 20 across the Elkhorn and its tributaries. Despite the numerous washouts that cut rail and road traffic through substantial parts of the upper basin, local observers didn't consider this set of floods particularly serious or damaging.[15] At Neligh, the Elkhorn flooded Riverside Park and washed out state highway 14, the city's major north-south road and an important route into town for those who lived south of the river. The road remained closed for more than a week.[16] Through the end of June, a half-dozen families near Beemer were evacuated from their homes and floodwaters damaged or closed railroad and highway bridges in most basin counties from Cuming upstream to Holt. Even with railroad bridges repaired, train traffic operated on slow orders through the end of the month due to dangerous conditions on the tracks. The cost and inconvenience of living with the Elkhorn extended to health concerns. State officials urged residents to get typhoid fever immunizations, to sterilize wells, and to discard or carefully cook produce that had come into contact with floodwaters. To fight an expected boom in the mosquito population, health officials urged residents to spray pools of standing water with a solution of creosote and kerosene or DDT.[17] Tallying the cost of the 1947 floods basinwide presents a challenge, but one county, Stanton, estimated damages just to roads and bridges at $85,000.[18]

The 1960s

After the 1950s passed in relative dryness for the floodplain, a cluster of floods in the 1960s reinvigorated dialogue between the Elkhorn and its people. The wet decade saw plenty of local action in the basin as a number of flood control projects came to fruition. The 1960s also witnessed new flood-related policy proposals nationally and the beginnings of a creative, ecologically sensitive bureaucratic approach to natural resource management on the state level. Collectively these events charted new directions in living with rivers.

Barely had the calendar turned to 1960 when ice jams brought flooding in March and rainstorms in June. The rainstorm flooding generated more attention starting on the night of June 9 when a downpour sent many Elkhorn tributaries on a "mild rampage." One such rampage knocked out the Highway 24 bridge severing traffic between Norfolk and Stanton.[19]

A stronger storm on June 19 hit the basin with damaging 69-mph winds and as much as seven inches of rain. The wind wrecked two hangars and damaged six airplanes at Norfolk's airport. The rain turned Elkhorn tributaries deadly. Near Madison a farmer lost thirty pigs that drowned in their pens. As usual, the floods washed out railroad tracks and bridges stopping or delaying traffic. On State Spur 24 between Stanton and Norfolk, Donald Henrichs, a 25-year-old farmer from Pilger, died when his car plunged into a washout. In all, the June 1960 storm washed out nine bridges in the basin and led to Madison and Dodge counties being designated disaster areas by the Small Business Administration.[20]

As costly and deadly as the 1960s floods were, they don't rank among the worst in the basin's history; yet they inspired a frustrated editorial in the *Norfolk Daily News* that decried wasteful damage and called for immediate action. The newspaper's editors catalogued the season's damages—nine bridges destroyed or badly damaged, houses and towns flooded, Norfolk's Ta-Ha-Zouka Park underwater, cropland damaged, thousands of tons of topsoil flushed down the Elkhorn, a motorist killed—and wrung their hands that the valley stood almost helpless in the face of flood threats. "Much of this could have been prevented by flood control," the editorial laments. The editorial reflects a keen awareness of the Elkhorn and its tributaries, and it reflects assumptions about the relationship between people and the river rooted in decisions and ideas brought to the basin generations earlier. It doesn't question the placement of crops, livestock, roads, railroads, settlements, dwellings, and recreational facilities in the floodplain. It assumes the power of the people to apply technology and knowledge to modify the floodplain to make it the domain of humans rather than the domain of the river. It assumes the proper use of a floodplain is to house valuable, productive assets. And in its conclusion, it reflects frustration with people dragging their feet as well as an acknowledgment that in 1960 people in northeast Nebraska continue to live in a kind of untamed frontier: "The time has come to stop the shameful waste and to control the wild waters."[21]

The decade's most problematic floods came in 1962 and 1966, and they reflected the mercurial nature of mid-continent weather patterns found in the Great Plains as well as the troubles related to occupying the Elkhorn's floodplain. An unusual confluence of precipitation events conspired in late March 1962. Winter left a thick blanket of snow covering the basin. A thaw began melting that snow over still-frozen soil, creating unusually high runoff because the land couldn't absorb much water. The thaw also broke the Elkhorn's ice, which began flowing downstream … as well as it could. Then substantial spring rain complicated the situation by hastening the snowmelt and by adding to the runoff headed for streams and rivers. As this unfolded, ice jams formed in the Elkhorn and some tributaries impeding stream flows and effectively damming portions and creating pockets of high water. For example, near Hooper an ice jam three-quarters of a mile long sent water into the town. The town's ball park sat under four to five feet of water. Workers frantically built a wall of sandbags at the city auditorium in an effort to save it from flood damage and others moved livestock to high ground. Residents of northeast Pierce had to evacuate, too, as the North Fork flooded their neighborhood.[22]

Four days into the flood, residents of northeast Norfolk got word that they should evacuate their homes as the North Fork swelled into its floodplain. South of the evacuated area, residents plugged their sewers as they tried to stop water backing up into their homes from the North Fork. At the junior college located in town, the president excused all male students from classes so they could join sandbagging brigades. In other parts of town, residents threw up earthen dikes as a defense against the rising water. At its height, the 1962 flood covered between 80 and 100 city blocks in Norfolk. The flood took a hefty toll, even if measured only at Norfolk, where damage estimates started at $550,000 and ran to $1 million.[23]

The 1962 flood also seems to have had a profound psychological impact on basin residents. It inspired memories of worst floods as people tried to make sense of the catastrophic damage. The *Norfolk Daily News* reflected the city's flood consciousness by noting that this was the worst flood since 1944. It also ran a retrospective piece on the flood of 1912, which happened to have hit on the same days. As the flood unfolded, the paper ran a story about flood control projects planned for Pierce and West Point that promised a drier 1963 for those cities. The story noted that those communities had taken steps to cooperate with the federal government in implementing flood control projects and that the federal budget proposed for the next fiscal year included money to build the projects. A high-ranking official with the Omaha Army Engineer District observed that similar projects proposed for Norfolk, Tilden, Battle Creek, Waterloo, Meadow Grove, Scribner, Hooper, and Winslow stood to qualify for federal funding if the locales would support them. Specifically, local governments had to provide rights-of-way for the projects, which Beemer and West Point had done. Whatever considerations had led the people of Norfolk to drag their feet in supporting a flood control project seemed to be dissipating rapidly in the aftermath of the 1962 flood. At the end of March, an outfit calling itself the Norfolk Action Committee issued a proclamation urging the city council to take the financial and legal steps necessary to implement the flood control plan recommended by the Corps of Engineers in 1947.[24]

The people of Hooper didn't need any more convincing. Two floods had hit Hooper in 1960. In March high water damaged the floor of the community's auditorium. Repairs had barely been completed when a June storm dumped nearly five inches of rain on the area. Workers quickly threw up sandbags around the auditorium and saved the repaired floor from further damage. Two years later a massive ice jam fifty yards thick and twelve- to fifteen-feet high blocked the Elkhorn and again sent water toward Hooper. Workers scrambled to throw up a dike at the village's west end—and again at the auditorium. Five sticks of dynamite blew a channel through the ice jam; and Hooperites claimed victory over the flood, according to a local history, but not before a number of families had to be evacuated.[25]

For years Hooperites had toyed with the notion of building a protective dike for the town, but it took the floods of the early 1960s to move them to act. In August 1963 village officials contacted the Corps of Engineers to begin planning a joint dike construction venture. The Corps designed an 11,000-foot dike that would protect Hooper from floods five feet higher than the 1944 flood of record. The US government would pay for most of the construction (about $266,000) and Hooper would pay a much smaller portion (about $34,000). Hooper held a special bond election on August 31, 1965, for the project. The local paper threw its weight behind the project. A few days before the election, the paper ran photographs of previous floods to support its argument that a "yes" vote would bring an end to sandbagging, waiting in suspense for flood warning sirens to sound during heavy rains or quick melts, evacuations, sewer backups, and the massive cleanups that followed the Elkhorn's all-too-regular floods. The Hooper Fire Department, Commercial Club, and Lion's Club all ran "Vote Yes" ads in the paper. Voters approved the bond overwhelmingly, 238–65. Work began the following week and the completed dike was turned over to Hooper on November 15, 1966. Months later, when heavy rains came in June 1967, the dike began paying dividends by diverting high water that locals estimated would have caused $40,000 damage to Hooper.[26]

Upstream, the people of Norfolk evidently needed one more push to embrace the Corps' plan. Rain, snowmelt, and ice jams in 1966 created a series of floods remarkably similar to those of 1962, but in early February before the calendar had even turned to spring. A warm spell with temperatures near 50 degrees probably felt good to basin residents in the middle of a Nebraska winter, but the warm weather came with a heavy rain—1.31 inches at Norfolk on February 8. The rain fell once again on melting snow and still-frozen ground, which sent runoff into streams and rivers. Ice blocked both the North Fork and the Elkhorn. Huge chunks of ice wedged against the First Street bridge over the Elkhorn south of Norfolk created a massive dam that threatened to flood much of the city's south side. A newspaper photo captured a car driving in axle-deep water on Omaha Avenue (US 275), proof that blasting with dynamite had failed to dislodge the ice dam. Although ice jams plagued the Elkhorn far upstream from Norfolk and promised continued drainage problems for the city, the North Fork managed flow adequate to prevent flooding. Elsewhere in northeast Nebraska, the combination of ice jams and storm runoff caused widespread lowland flooding.[27]

Downstream at Wisner, long-time residents ranked the flash flood there among the community's worst. Families and livestock had to be evacuated from lowlands and one livestock operation lost more than 200 head of cattle. Homes on the southwest side of town near the railroad tracks had two to three feet of water in them. Farther downstream, the Elkhorn washed out a railroad bridge between Scribner and West Point, which halted rail service southeast from Norfolk for several days.[28]

Every Norfolk city employee turned out to fight the flood with assistance from the National Guard. Firefighters evacuated people and livestock from the city's extreme south side, which fronts the Elkhorn. In that neighborhood Ta-Ha-Zouka Park, considered a local showplace, suffered damage to drinking fountains, picnic tables, and shelter houses. In the park's zoo, a fox, coyote, rabbits, and other animals perished in the flood. The bears, which climbed atop picnic tables in their enclosure, survived. When the ice jam, which didn't budge during blasting, broke on its own on February 9, the Elkhorn drained the Norfolk area but created problems downstream, notably east of West Point where high water closed US 275. Ice jams created additional flooding all the way to the Elkhorn's mouth, where a huge jam sent water over thousands of acres of farmland. The high water also undermined the approaches to two Interstate 80 bridges over a drainage channel. The State Department of Roads closed the westbound I-80 bridge, disrupting traffic on the newest transportation artery placed in the Elkhorn's floodplain. A bit of good news, though, came at the site of much previous flood damage from the North Fork: Pierce, behind new flood walls, was dry.[29]

As the basin's streams returned to normal, observers raised calls, again, for change. The *Norfolk Daily News* told its readers that they should have more conversations about flood control. Furthermore, they should support the city's cooperation with the Corps of Engineers in completing the flood control projects recommended in 1947 as quickly as possible. Governor Frank Morrison visited Norfolk and other spots in northeast Nebraska to survey flood damage. What he saw made him lament the massive loss of topsoil to flood-related erosion and inspired him to call for more aggressive soil conservation to protect the area's valuable soil as well as to minimize storm runoff.[30] Whatever resistance or apathy had delayed progress on the Norfolk flood channel disappeared, perhaps carried off metaphorically by the receding flood, and construction began within months. By the end of 1968, Norfolk had its flood channel.

Norfolk residents didn't have to wait long for the channel to pay dividends. A thunderstorm dumped more than four inches of rain—nearly what would fall in the entire month of June—on the area during the night of June 24–25, 1969. On the south side, flash floods closed Omaha Avenue and US 81, east-west and north-south thoroughfares, respectively, for several hours. West and northwest Norfolk also experienced some flooding. But downtown reported no problems. The new flood control bypass channel that ran along Norfolk's north and east had much water and was "carrying it very well," according to Norfolk City Street Commissioner Earl Smith.[31] But even as the new flood channel did its job, the 1969 storm showed that old problems of living with the Elkhorn persisted. The high water had created a large log jam at the South First Street bridge. In this case old logs, whole trees, and other debris didn't threaten to harm the bridge, but they did cause water to back up and to make nearby lowland flooding worse than it would have been without the jam. Workers tried to blast the jam apart with dynamite, but that failed and a crane had to be used to pull apart the blockage.[32] To this point, assumptions underlying how to live with the Elkhorn had changed little since the late 1800s. People still sought to impose their will on the river, to improve or perfect it, even as the scale of modifications and the resources available to enact them increased and the cast of characters expanded to include players from outside the basin, namely federal agencies. But new ideas were circulating in Lincoln and would soon be on their way to the Elkhorn.

As the floods of the 1960s played out in the Elkhorn Basin, and elsewhere in Nebraska it should be noted, a new orientation toward living with rivers was developing in offices of the state government in Lincoln. In 1963 Nebraska established the Small Watersheds Flood Control Fund. This revolving state fund provided money to help sponsors, which might be cities, counties, or natural resources districts, obtain land rights necessary to build flood abatement projects. The program required sponsors to think broadly about flood abatement rather than to focus narrowly on small-scale projects. For example, a sponsor had to demonstrate that at least seventy-five percent of land upstream from a proposed flood abatement project would be treated to manage drainage and runoff. The program also contained measures to take flood prone land out of circulation. On completion of an abatement project, land acquired for the project had to be sold, with local governments getting first chance to buy the property for public use. Maintaining such property as undeveloped green space would provide beneficial recreation facilities as well as reduce potential flood damage by prohibiting construction. If no public entity showed interest, the land could be sold to a private buyer with a conservation easement attached. Proceeds from land sales returned to the fund for use in other projects. This program supported at least two projects in the Elkhorn Basin—the Pilger Watershed, 118 acres northeast of Pilger, and Corporation Gulch, more than 107 acres on the south side of Norfolk.[33]

In 1965 the state legislature ordered the Nebraska Soil and Water Conservation Commission to draft a "comprehensive program of resource development, conservation, and utilization for the soil and water of this State in cooperation with other local, state and federal agencies and organizations." [34] In submitting the plan, Robert M. Bell, the commission's chairman, acknowledged what some had surely known all along, "until the day that man can control nature, he must expect the river to occasionally assert its right to the valley bottoms."[35] The forward looking plan called for integrated policies on flood control, irrigation, pollution, land use, and more to be planned statewide.[36]

Although the plan offered what might be considered a progressive "the river will sometimes have its way" outlook on its first page, the flood prevention plan on page two sounded familiar notes of "prevention is possible." In fact, the flood prevention section opens with the bold statement, "The prevention of floods has been a dream of man since time began. Only by the prevention of floods can our villages and cities be secure and our fertile valleys be farmed." But, alongside that old-fashioned sounding sentiment, the plan prescribes a "total program" for treating watersheds in at least six integrated ways: soil and water conservation treatment on the uplands; detention structures on tributaries; main stream structures storing large amounts of floodwater; channel improvement; proper land use in flood hazard areas to reduce damages where other means of control are not utilized or not sufficient; and, effective emergency action.[37]

Planners based this approach partly on reports that despite the federal government spending more than $7 billion on flood control between 1936 and 1962, flood damages increased each year. Researchers estimated that 25 percent of those increased damages came from an increase in the amount of flooding; 30 percent came from an increase in building and other uses of floodplains; and, 45 percent could be attributed to rising property values. In other words, there were more floods, but 75 percent of increasing damages involved rising value of floodplain real estate, the value of what had already been placed in floodplains, and new things being placed in floodplains. The state's multi-pronged approach couldn't stop flooding or undo existing floodplain development, but it suggested methods for controlling drainage and it could manage what people could put into a floodplain.[38]

The plan's recommendations for land use regulation sound particularly sensible and probably had the greatest potential to generate controversy. Land use regulation involved identifying areas "susceptible to flooding and limiting their uses to those which will not be seriously damaged or present a hazard to life if flooded." Put another way, it tries to prevent mistakes in floodplain development. Among other things, it attempts to prevent the erection of structures in the floodplain that might limit channel capacity and increase flood heights, to prevent loss of life, to reduce public expenditures on things such as rescuing people who have unwisely built in a floodplain, and, by limiting the placement of valuable assets in a floodplain reduces demand for additional expenditures, public or private, on reservoirs, levees, and other structures designed to protect things that should not be in a floodplain in the first place. In other words, land use planning represented an attempt to learn from the history of floods chronicled in the book you hold in your hands and to avoid repeating mistakes of earlier generations.[39]

Logical as those ideas might sound, local authorities had taken few steps to implement them and had struggled to implement land use regulations uniformly. The state plan aimed to help local governments not only by providing ideas and examples of land use regulations, but also by placing responsibility at the state level, which would allow coherent, uniform regulation and take the heat of opposition off local government officials who might have trouble saying "No" to a neighbor's request to build unwisely in a floodplain. The plan also offered helpful hints for answering critics who claimed land use regulation infringed on their property rights. The individual property right argument, for example, often neglects the fact that one owner's unwise floodplain development has harmful effects on neighboring properties. And it overlooks the financial burden created for the wider public by tax dollars spent on flood fighting, flood relief, and structural flood control.[40]

The water plan also presented a priority list of communities that would most benefit from floodplain zoning. To set the list, the commission used flood damages recorded by the Corps of Engineers from 1940 through 1965 (records before 1940 were too fragmentary to use) and a community's population growth during the same period. Communities that demonstrated the greatest flood damages and the greatest population growth went to the top of the list. By this calculation, Lincoln sat atop the list. Norfolk ranked fourth, thanks largely to the massive losses it had suffered from 1940 through 1965. With recorded damage of $12.2 million, Norfolk was by far the most flood-damaged community in Nebraska during the study period. In fact, three of the top seven most damaged communities (Norfolk, Scribner, and West Point) were in the Elkhorn Basin.[41]

The report evidently impressed members of Nebraska's unicameral legislature. Shortly after receiving the document, they passed "Nebraska's Flood Plain Management Act" by unanimous vote. Pursuing the plan's recommendations, especially those aimed at coordinating various local and state agencies in service of land use regulation, put Nebraskans including the sons and daughters of the Elkhorn on new paths toward living with their rivers.[42] Assuming Nebraska sits entirely within the Great Plains, this by definition pointed to new ways of living with Plains rivers.

As Nebraska reoriented its approach to living with rivers and flooding, so did the nation and at virtually the same time. Congress expanded federal economic assistance for flood victims with the National Flood Insurance Act of 1968. This law created the National Flood Insurance Program, a joint venture between the government and private insurance companies, with two goals. First, Congress hoped to promote "rational use of the floodplain." Second, Congress hoped to encourage better funding of disaster relief by getting more floodplain occupants to pay insurance premiums before suffering damages. The government would use private insurers to offer below-cost flood insurance rates, but only to people in communities that adopted land-use restrictions and other regulations designed to minimize flood damage. To be eligible, communities had to adopt zoning restrictions, building standards, flood-proofing, and emergency preparedness plans. Congress passed additional measures in 1973 (Flood Disaster Protection Act) and 1988 (Stafford Disaster Relief and Emergency Assistance Amendments) to strengthen this agenda.[43] These new policies reflected new policy approaches and motivated new ways of living with rivers that eventually trickled down to the Elkhorn Basin.

Meanwhile, back in Lincoln, concern about flooding and the confusing, overlapping web of agencies charged with handling some aspect of resource management brought forth a new entity unique to Nebraska, the natural resource district. As of 1969, Nebraska contained hundreds of special purpose districts such as watershed boards and county soil and water conservation districts. State Senator Maurice Kremer sought to simplify and rationalize that system by introducing legislation in 1969 that would create a new system of natural resource districts (NRDs) that would begin operation in 1972. The original 24 NRDs (reduced to 23 on the merger of two districts in 1989) were given jurisdictions with boundaries defined by the state's major river basins (Map 5-1). This would allow them to develop management practices appropriate for similar topography within a river basin. The NRDs are local government units with elected directors funded by a portion of the district's property taxes. Kremer and his supporters viewed this structure as an ideal combination of a broad watershed approach to conservation administered by locally-elected officials. Although the NRDs were founded to address flood control, soil erosion, irrigation runoff, and groundwater supply and quality problems, they have evolved to address a dozen responsibilities. NRD flood control projects include dams, terraces, drainage ditches, and reservoirs. The Elkhorn flows through two NRDs, the Upper Elkhorn NRD (mostly Rock, Holt, and Antelope counties) and the Lower Elkhorn NRD (Pierce and Madison counties to the river's mouth).[44] These new agencies and programs pointed to new ways of living with rivers, which eventually found their way to places like the Elkhorn Basin. But old habits died hard and some people changed only grudgingly, even as the floods kept coming.

The 1970s

The Elkhorn Basin felt little flooding during the 1970s, but the legacy of the 1960s could be detected by those paying careful attention. For example, a late winter flood hit the Hooper area on February 25, 1971. A blizzard had left thick snow cover in the preceding week and ice jams on the Elkhorn and its tributaries blocked drainage. Heavy chunks of ice slammed into bridges causing damage and constructing makeshift dams that sent water onto thousands of acres of lowland farms. Hooper escaped damage, though, thanks to its dike. The town's centennial history celebrated the structure noting it "again proved its worth stopping floodwaters which rolled down from the west."[45] The spring flood of 1978, felt most keenly along the Platte and at the Elkhorn's mouth, highlighted those legacies. At Norfolk, high water in the North Fork flowed as planned through the flood diversion channel. The stream's original channel, which runs through downtown Norfolk, ran full but did not overflow. The *Norfolk Daily News* noted that in years before the diversion channel's completion, flows similar to those in 1978 would likely have spilled into downtown. This time, as had happened the year after the channel's completion, the channel easily took would-be floodwaters around the city to its north and east. Downtown stayed dry. Near the confluence of the Elkhorn and the Platte, especially at Valley (in the Platte's valley), hundreds of people had to evacuate their homes. After surveying flood damage in the region, Governor James Exon noted that some of the appalling damage could be attributed to the large number of homes built recently in the floodplains.

Exon's comment suggested that even in the late twentieth century people had not learned lessons from more than a century of living with the region's rivers. It also suggested that a phenomenon playing out elsewhere in the country was happening in the Elkhorn Basin, too. With levees, dikes, and other

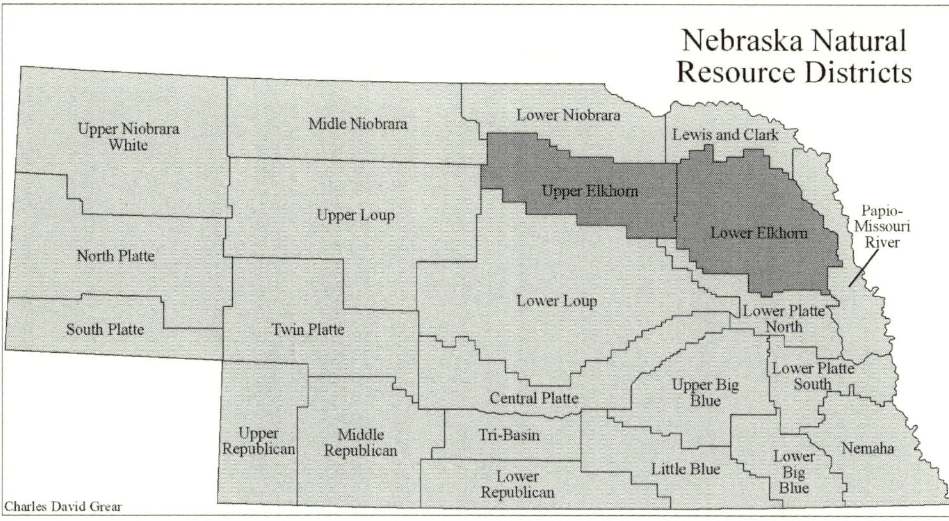

Map 5-1: Nebraska's Natural Resource Districts were organized according to river basin boundaries rather than county lines. Map by Charles David Grear.

flood control structures in place and the advent of flood insurance, people seem to have been lulled into thinking it was appropriate or safe to build in the floodplain. For the Elkhorn region, the flood of 1978 proved that wasn't the case. Furthermore, an editorial in the *Norfolk Daily News* points to the sense of complacency, or maybe greed, that exacerbated the flood's damage despite the existence of measures designed to mitigate it. Few homeowners in the areas of greatest damage had bothered to purchase flood insurance. They would have to rely on charitable organizations such as the Red Cross to help them recover from the flood. Whatever aid came from such sources, though, would fall short of fully restoring losses in many cases. The paper made it plain that Norfolkans and others who lived in river valleys ought to learn from the experience: where flood insurance was available, people should buy it.[46]

The 1980s–1990s

The decade of the 1980s and the first few years of the 1990s were relatively peaceful with respect to Elkhorn flooding; nevertheless, some voices raised to sing familiar refrains. June 1984 brought damaging flooding by the Missouri and Elkhorn rivers to seven counties in eastern Nebraska. The floods weren't of a magnitude to register among either basin's most significant, but the damage and the responses demonstrated continuity in the area's hydraulic history. Transportation suffered as floodwaters caused more than $600,000 damage to state-maintained bridges and roads in the area. The Elkhorn damaged six state bridges.[47]

The flood also eroded land and forced people from homes in the floodplain. The Lind family, which owned property in the Elkhorn Basin outside Valley, Nebraska, dating to the 1950s suffered both erosion and evacuation. In 1954, the family's spread covered 117 acres. Decades of erosion from spring flooding had reduced the plot to 75 acres by 1984. "The river keeps taking it away from us," Debby Ten Eyck, a Lind family member, told a newspaper reporter. "But we're OK. The farmers up river are losing their land and losing their livelihood. It's just terrible, that's all…Just think about all the river families going through this." Bud Lind, who lived on the tract at the time, added, "It's a mean river." Lind family members had to move out of two of the four homes on their land. They joined 675 other families across fourteen eastern Nebraska counties forced from their homes that June. As it had done for decades, the Red Cross provided charitable relief for flood victims. In this case, the Red Cross collected monetary donations from individuals and corporations to provide meals and other assistance to the flooded.[48]

Some evacuated families had problems beyond the scope of Red Cross assistance. At Waterloo, not far downstream from the Linds, Alvin and Gertrude French had to leave their home, a house trailer sitting on blocks about fifty feet from the Elkhorn's bank, when the flood brought about a foot of water into the trailer. The Frenches had lived in the spot for a dozen years and had never felt threatened by the Elkhorn. "For twelve springs and summers, we sat here and watched…big trees float down this river," Gertrude French told a newspaper reporter. "We never did think it would do this. These are things that happen to other people."[49]

When the Frenches returned from their evacuation, they found little damage to their trailer home. But, they found the Elkhorn had carved out a big chunk of the land that had separated their home from the river. After the flood, about three feet of sandy soil stood between their home and a steep drop to the water. A toolshed teetered on the edge, too. The flood left the Frenches to figure out how to relocate their home before it fell into the river. The retired couple who lived on Social Security income of $700 a month worried that they couldn't afford to move the house.[50]

The French's story, and the Lind's for that matter, show that people in the late twentieth century could and would ignore decades of flood history to move onto the Elkhorn's floodplain. The Frenches enjoyed the recreational benefits of living near the river, but their willingness to live that close to a river and to think that it couldn't harm them reflects a misplaced confidence in their predecessors' efforts to tame the river. Or maybe it suggests arrogance or ignorance. Their experience also shows that inexpensive real estate in a floodplain can be a draw for people on low or fixed incomes along a minor Nebraska river every bit as much as it can along the Missouri or the Mississippi.[51]

A full week of rain in 1990 brought a round of June flooding to the lower Elkhorn that drew comparisons to the floods of 1984 and 1944 and added evidence to support the conclusion that faith in the Elkhorn's taming had been misplaced. Mary Ann Fauss of Nickerson, whose property sat roughly equidistant from Maple Creek and the Elkhorn surveyed water five- to six-feet deep on farm fields and told a newspaper reporter, "It's higher than it's been here since 1944." A National Weather Service hydrologist compared the flood to the more modest one of 1984. Wherever the 1990 flood ranked, the Elkhorn came within six inches of topping a Washington County dike that separated the river from a mobile home park that housed sixty families.[52] Early in the basin's history it made sense for people to place some structures, namely mills, so close to the river. But in the years since the culture's divorce from the river's motive power, it made little sense to put homes in the floodplain. Here the Elkhorn echoed developments in other river's floodplains. The dike afforded a sense of security and seemed to say that some portion of the floodplain had become suitable for habitation. The presence there of mobile homes, though, suggests it was inexpensive real estate and being inhabited by people priced out of more secure, more reliably dry land.

Most towns and villages in the lower Elkhorn faced flood damage and from Arlington to Winslow, where the town of 150 had 300 volunteers stacking sandbags seven deep to raise a "dike" created by a roadbed. High water closed many roads within reach of rivers and streams in northeast Nebraska. Farmers who worked bottomlands saw their fields ruined and faced the prospect of replanting or even losing some production for the year, as well as the loss of topsoil. Dozens of floodplain residents had to evacuate their homes. At emergency shelters at churches they found food and other aid from the Red Cross and the Salvation Army, a pattern of charity that had become familiar over the years. Costly as those problems were, the flood's most heartbreaking story of damage came from rural West Point. Andrew Goeken, 77, who walked with the help of a four-legged walker, apparently went outside his house to check on his property and was swept away in water that reached five-feet deep in the vicinity of his home. Goeken's son found the body days later tangled in a fence more than a block from his home.[53]

Three years later the basin experienced another double-flood year with ice jams and snowmelt causing the problems in March and runoff from heavy rain bringing high water in June. The water that caused the March flooding fell to the basin in February when parts of northeast Nebraska got up to ten inches of snow. A ten-day warm spell in early March melted the snow and sent it rapidly into rivers choked with ice. Some of the most serious drainage problems happened on the lower Elkhorn and in the Platte at the two rivers' confluence where a five-foot thick ice jam extended two miles upstream. The runoff couldn't flow downstream and, instead, went over the rivers' banks. Similar problems in other eastern Nebraska river systems forced about 1,400 Nebraskans to evacuate their homes for several days, with several hundred of those evacuees in the Elkhorn basin and most of those from the West Point area. Both the Platte and the Elkhorn carved new channels in places, adding to the normal headaches associated with floods.[54]

About 375 people evacuated roughly 200 homes in West Point's Third Ward on Tuesday, March 9, when the Elkhorn climbed over a dike there. The IBP beef plant in town closed early so employees who lived in the threatened area could deal with the flood. An ice jam just downstream from West Point broke that afternoon, which ended the flood threat at West Point before the high water could damage homes. Even as the water receded at West Point, the Elkhorn closed major roads out of the city and West Point's Grace Lutheran Church held a prayer service for the flooded. Most evacuees were back in their homes by supper time that night, but a rush of water headed for downstream communities and the Platte, where it would cause trouble for others.[55]

The plant closing, the prayer service, and other responses to the 1993 flood show how high water could influence social and economic life in the floodplain. Other examples included the formation of a sandbagging crew that threw up a dike around West Point's sewage treatment plant to prevent floodwaters from washing sewage into the town and downstream. In addition to offering a prayer service, Grace Lutheran Church provided meals to evacuees. Other area churches also offered meals and use of their phones to the flooded. The Red Cross, as usual in Elkhorn floods, chipped in with help. "It's really interesting how anxious people are to volunteer in a small community," Dean

Lauritzen, Cuming County Red Cross disaster chairman, told a newspaper reporter. "You almost have more than you need. Businesses contributed food, and there was a great community spirit."[56] The floods brought people together in a time of crisis. Living with the Elkhorn brought out the best in people and exposed the weave of the community's social fabric in 1993 as it had a century earlier.

Community bonds got another boost just months later when a rainy June, including a storm that dropped nearly five inches of rain overnight on the Pilger area, brought flash flooding from the Elkhorn basin as far east as Des Moines. Families in Madison fled their homes ahead of the high water. At Pilger, at least six homes were evacuated. Elsewhere in the basin the storm brought the usual litany of problems ranging from erosion to the destruction of at least one bridge. The storm also prompted reflection on successful flood control measures that mitigated damage from the intense runoff. The Pilger Watershed project, for example, seemed to have minimized damage in the 4,700 acres it served. The Soil Conservation Service, in partnership with other state agencies and with federal funding assistance, built two dams to capture runoff. Motivated by damage to thirty-one homes from a five-inch rainfall in 1958, the first structure was built in 1959 and the second in the early 1960s. According to a local conservation official, the 1993 storm showed the structures working exactly as planned. A newspaper report supported his contention—the approximately five-inch rain of 1993 led to six or seven homes being evacuated, compared to the thirty-one homes that were damaged in 1958 and which sparked the construction of the projects. Nearby Norfolk suffered "minor" water damage, mostly in the form of flooded basements, despite a month of nearly record-breaking rainfall. That outcome prompted a feature story in the *Norfolk Daily News* celebrating the success of the city's flood control channel. Since the channel's construction, the paper noted, the city had not experienced any extensive flooding. Norfolk's city administrator told a reporter that no other construction project in the city's history had done more to protect property.[57]

The 1993 floods inspired two modification projects with a combined cost of more than $500,000 in 1994 near the Elkhorn's mouth outside Omaha. A $238,745 bank stabilization project aimed to halt erosion that threatened roads would be paid for by the US Soil Conservation District, the Papio-Missouri River Natural Resource District (NRD), and Sarpy County. Just upstream the Papio-Missouri NRD joined the Lower Platte North NRD and the federal government to pay about $280,000 to build up a low spot in the Elkhorn's bank to make it harder for floodwaters to spill onto farmland. High water came through the low spot in 1990, 1991, and in 1993 to cover thousands of acres of farmland and to damage roads and private property. The federal government paid for 75 percent of the projects' cost.[58]

Conclusion

The decades after the record floods of 1944 showed both continuity and change in the relationship between humans and the landscape of the Elkhorn Basin. The Corps of Engineers' plans inspired by the massive 1944 floods showed glimmers of a new era insofar as their idealized versions advocated a version of treating the basin holistically. The hard calculations of economic feasibility ruled out implementation of such a vision and even the Corps focused mostly on perfecting the landscape, but only in and around the river. That vision dominated through the 1960s.

In the 1960s and 1970s, as the Elkhorn continued to do what it does, the human side of the conversation moved into new territory. The adoption of a state water plan, which advocated an even wider holistic vision of watershed management than that articulated by the Corps, and the creation of natural resource districts, political institutions organized by watershed, showed considerable changes in how humans would live with and try to manage water, land, and the meeting of the twain. After decades in operation, the natural resource districts came to be seen as an improvement over the 154 assorted types of special districts they replaced. They streamlined an unwieldy bureaucracy, followed an ecologically sensible river basin organization that promoted rational basinwide approaches to water management, and enjoyed reliable funding based on property taxes that supported sustained programs.[59]

Concurrent changes in federal policy, namely the National Flood Insurance Program, also shaped these policies. People seemed to have realized, finally, that their ancestors had overstepped their place and that trying to bend the landscape and the hydrological system of the Elkhorn River to human will was a losing proposition. Instead, at least some of the humans involved in the story came to the conclusion that a better way to live with the Elkhorn would include changing how people behaved by keeping humans and their structures and other assets out of the floodplain. People would still keep trying to perfect the landscape by managing runoff and helping the Elkhorn move water downriver as efficiently as possible, but with a little humility people would limit their incursions into the contested terrain of the floodplain. Fully executing that vision, though, will likely be an upstream journey. It would mean repudiating the legacy of the pioneers who staked claims in the floodplain, compromising or giving up on some riparian property rights, and retreating from the notion that people can perfect the landscape to accommodate inhabitation of the floodplain. The new version of a perfected floodplain would be one with as little human presence as possible.

Conclusion

TAKING OUT: FLOODING AND CLIMATE CHANGE ON THE PLAINS IN THE TWENTY-FIRST CENTURY

"Combatting Nature and high water is a perpetual battle that is seldom entirely won."

H. H. Tracy, City Engineer, Norfolk[1]

"People never master their environment; they bargain with it."

Elliott West in *The Contested Plains*[2]

More than a decade into the twenty-first century, the Elkhorn continues to flow through the souls of its sons and daughters and to provide a focal point for activity in its basin. Take, for example, the fate of a recreational tubing expedition in June 2014. A group of tubers set out on Father's Day to make the most of a beautiful late spring day and one of the great recreational assets at hand on the Plains: a gently flowing river. The tubers set out for an easy float on the lower Elkhorn near Waterloo. When the group disappeared around a bend and didn't re-emerge, an observer on nearby land called first responders for help. Fortunately, emergency workers found the tubers stranded on a sandbar and brought them to safety on the rescue squad's air boat.

The tubers hadn't been randomly stranded; the river had risen rapidly, seven feet in a matter of hours according to a television news report, and the high flow and quickly moving floating debris presented mortal danger that hadn't existed when the tubers began their journey. Rainstorms north of Waterloo on Saturday night had created the hazard, which took until mid-day Sunday to catch the tubers, who *thought* they had carefully checked the river before embarking, off guard. Recreation and peril both remain hallmarks of living with the Elkhorn.[3]

The Twenty-First Century

The first decade of the twenty-first century brought a respite from major flooding in most of the Elkhorn Basin, but high water never strayed far from people's minds and the dialogue between humans and the river continued. Federal policy, for example, affected land use in the floodplain. The city of Norfolk managed part of its floodplain by buying portions through a program sponsored by a Federal Emergency Management Agency (FEMA) grant. The city identified thirty-three parcels in the floodplain and late in 2000 applied for a FEMA grant to offset the cost of acquisition. FEMA awarded the grant, which covered 75 percent of the purchase price, in 2001; and the city won additional funds from the Lower Elkhorn Natural Resources District and the Norfolk Sanitary District. Once the city gained control, city workers removed brush, discarded tires and appliances, and disposed of barrels of trash. Those things had not only made the land an eyesore, but also had obstructed the flow of high water. Accepting the federal grant also meant accepting restrictions on subsequent land use in the floodplain. Only open-space projects, parking lots or recreational projects such as hiking and biking trails, for example, could be built on the floodplain. This policy reflected a shift away from trying to overpower the river to stop or mitigate flooding to an approach that more or less acknowledged the inevitability of flooding, tried to limit damage by restricting floodplain development, and attempted to ease flooding by removing obstacles to drainage.

The policy also illustrated the difficulty of undoing generations of floodplain development. In the project's early years, eleven of the thirty-three landowners were willing to participate in the program. By the end of 2003, the city had acquired five parcels and sought to use eminent domain to obtain a sixth. Other efforts by FEMA to restrict floodplain development met hostile responses from landowners who believed the regulations would harm them financially by requiring them to buy flood insurance or making them ineligible for flood insurance. Soil-conservation-as-flood-control remained part of the conversation, too, as agriculture specialists continued to try to convince farmers to adopt practices that would allow the land to hold more water and slow runoff. One estimated that farmers in the 35,000-acre Battle Creek watershed alone, by adopting a type of no-till farming that would allow land to absorb more water, could keep nearly a billion gallons of runoff out of Battle Creek.[4]

Near the confluence of the Elkhorn and the Platte, Dial Land Development Corporation brought an unusual variation to floodplain utilization when it located an upscale housing development in Waterloo, which sits between the two rivers. A more common problem since World War II has been the placement of low-income housing, notably trailer parks, on floodplain land protected by dikes or levees. In Waterloo, the Homes at River Road subdivision put more than a hundred homes valued from $160,000 to $220,000 on land protected from flooding by a levee that surrounded most of the town. Building behind a levee presented an obvious issue, but not the only one. Grading for the development created a low spot in a corner of the development adjacent to the levee. A heavy rain or snowmelt left the area filled with water. The developer intended to use the spot as a water retention area and, possibly, a wetland. The Waterloo village board, however, worried that the spot might cause flooding or, worse, damage the levee. The developer cited the USACEs' approval of the project as proof that the risk of flooding and levee damage was minimal. The village, having extensive experience with flooding, proposed that the developer sign an indemnification agreement assuming responsibility for damages to the levee or other property if flooding occurred. After months of negotiation, the development's homeowners association and developer agreed to assume liability for any injury or damage from water on the site with financial coverage provided by a $5 million insurance policy.[5]

Later in the decade, local government officials responsible for administering the low-lying land of the Elkhorn and Platte valleys prepared for flood season with action both sophisticated and jaded. When conditions in late February 2007 pointed to the likelihood of flooding, officials asked people living along the rivers to identify an evacuation destination in the event they had to flee high water. They also asked people with cabins or summer homes in flood-prone areas to remove valuables before water rose. This sensible advice reflected longstanding experience with flooding and at the same time showed that despite longstanding experience with flooding people still lived on the floodplain, which seems to defy common sense. Officials assured residents that they would be able to assist with evacuations, should they be necessary, but they also advised them to be prepared to sign a document acknowledging evacuation. During past floods government agencies found themselves rescuing people who had been evacuated but who had returned to their property to retrieve valuables only to become stranded. In other words, the local governments assisted with evacuation and then got stuck rescuing people who returned home before the danger had passed. Not anymore. In signing the evacuation document, evacuees acknowledged that if they returned to their property and needed to be rescued, they would be billed for the expense.[6]

Despite a century-and-a-half of experience and myriad efforts to tame and contain the Elkhorn, basin residents found themselves at the river's mercy again in 2010. That year's flood displaced 1944's as the basin's worst and showed how vulnerable people remained to high water. Even in 2010, sons and daughters of the Elkhorn were attuned to the watery threats that accompanied spring's arrival. Melting snow and spring rains brought *yearly* fears of flooding to much of eastern Nebraska. So it was in early March 2010 when the National Weather Service predicted minor to moderate flooding for the Elkhorn Basin. The ice-melt part of that year's flood season passed as predicted, but thunderstorm season in June brought big problems and the basin's worst flooding ever.[7]

The worst-ever flood started with early June rainstorms that saturated the ground in substantial portions of the basin. Early signs of trouble came from near the Elkhorn's mouth where the state closed two riverside recreation areas on June 8 due to high water. Flood forecasters predicted campers would be out of luck at those sites for at least a week. "We're in a pretty darn busy weather pattern right now," a meteorologist said in what turned out to be a dramatic understatement. In the next two days problems moved upstream as the Elkhorn and its tributaries flooded highways, roads, and farmland in Dodge and Washington counties and as far upstream as Stanton. The people of Nickerson were ready. They had prepared a stockpile of sandbags, which they threw up to protect themselves from Maple Creek as it rose to flood stage.[8]

At about the same time heavy rains in the upper Elkhorn Basin, especially in Holt County, meant that runoff there would soon be compounding problems downstream. Water covered virtually all of Inman, a Holt County village. With water covering many of the roads leading out of O'Neill, the county's largest city had become an island as of Saturday, June 12. Ranchers went out on horseback to move cattle to high ground as county officials predicted the Elkhorn wouldn't crest until the following Tuesday and that sandbagging wouldn't help much. The river hit 12.2 feet at Ewing on Sunday, passing the previous record high stage of 11.3 feet set in 1947, and it was still rising. In terms of stream flow, the Elkhorn hit 28,000 cubic feet per second, triple the record set in 1947. Deb Hilker, Holt County emergency management coordinator, told a newspaper reporter that the Elkhorn "looks like a lake more than a river."[9] That "lake" was headed downstream.

The next counties downstream, Antelope and Madison, dealt not only with the water that fell on them but also with the drainage from Holt County. At

Clearwater, the first community on the Elkhorn inside the Antelope County line, residents worked from Sunday night into Monday's early morning hours to build a sandbag wall between their homes and businesses and the swelling Elkhorn. Hundreds of volunteers lugged and stacked sandbags. Some worked the line for eighteen hours. Others worked with a Caterpillar bulldozer to shore up weak spots. The rising Elkhorn probed the berm unceasingly until it found a gap, which it opened into a three-by-eight-foot hole. Water rushed into the town carrying sixty- to ninety-pound sandbags past volunteers who had hopefully stacked them against the river. It looked like the ground was floating. The fire chief ordered the north half of the Clearwater evacuated. Volunteers rallied to fill the gap, and they stayed with the fight for another forty-four hours. Clearwater survived, but the Elkhorn damaged about a hundred homes.[10]

At Neligh, the Elkhorn showed yet again the perils of occupying the floodplain. The river reached into the city's park, but did its most thorough damage to bridges. It swept trees into the piers of the mill bridge, which had been erected as a vital link for farmers who needed access to the mill and markets, and caused it to collapse. The bridge, which holds a spot on the National Register of Historic Places, had withstood many floods but succumbed to the rise of 2010. The high water also washed out part of the south approach in such a way that the bridge no longer connected the two banks. To become functional, a jump span would have to be added to connect the original bridge to the "new" south bank. Not far downstream, the river ripped out a swinging bridge that had connected Riverside Park to the opposite bank since 1935. According to local lore, the bridge's builder assured city officials the structure would stand for fifty years. It lasted seventy-five, during which it withstood some of the river's worst floods. The flood of 2010 proved too much, but left behind a pair of twisted steel I-beam supports that provide vivid, mute testimony to that flood's destructive power.[11]

The river also hurled one more insult at the Neligh Mill, which had become a state historic site and museum. Floodwaters packed the eleven-foot deep flume with sand, damaged the penstock, and washed away part of the mill dam. The river washed out supports for a suspension bridge that dated to the mill's early days and toppled sixty-foot cottonwood trees on the bank. The site closed for four days during the flood. "Normally, there's two to three feet of water [in the river]," Don "Harv" Ofe, curator of the Neligh Mill State Historic Site, told a reporter. "During the flood, there was fifteen feet of water. And it stayed up for three days." Repair work began in December with grants from the Federal Emergency Management Agency covering 75 percent of the cost and the Nebraska Emergency Management Agency and the Nebraska State Historical Society splitting the balance.[12]

The Elkhorn's "slow march of destruction" carried uprooted trees and other debris downstream into the Norfolk area where it took its most tragic toll. The Elkhorn had become something never before recorded or, likely, imagined. On Tuesday, the Elkhorn's flow near Norfolk registered 38,800 cubic feet per second. To put that in perspective, the flow rate for a "100-year flood," or a flood of a magnitude that would be expected to occur once every one hundred years, would be 30,000 cubic feet per second. Or, to put it another way, for the five-year period from 2010 through 2015, the Elkhorn's peak annual flow, which normally occurs in the spring or early summer, never exceeded 10,000 cubic feet per second. In four of those years it peaked at about 1,100 cubic feet per second. The remarkable 2010 flood fell into the 100- to 500-year flood category.[13]

Swollen to unprecedented proportions, the Elkhorn brought spectacular force to bear on the Nebraska Central Railroad bridge spanning the river southwest of Norfolk. The bridge collapsed into the river just before 1 p.m. on Tuesday, June 15. The collapse dumped three railroad employees, who had been on the bridge inspecting it for flood damage, into the Elkhorn's dark, roiling water. Two of the men escaped unharmed, but the third, Jeffery Scholl, aged thirty-eight of Norfolk, disappeared. Rescuers searched for Scholl by air through the rest of the day. As night fell, Norfolk issued a mandatory evacuation of its south-side neighborhoods near the river, which forecasters predicted would crest at an inch higher than the levee on its banks, and rescuers abandoned their search for the railroad worker.[14]

Rescuers located Scholl's body on Thursday evening, more than two full days after the collapse. Scholl, who worked for the Nebraska Central as a train engineer for more than ten years, was survived by his wife, Suzie, and two sons, ages six and three.[15]

The bridge collapse not only cost a life and devastated the Scholl family and friends, but also contributed to the economic cost of the 2010 flood. The Nebraska Central provided critical transportation for several Norfolk companies, including Nucor Steel, Louis Dreyfus Commodities ethanol plant, and Norfolk Iron and Metal. The Nebraska governor's office estimated that the railroad's service to businesses indirectly affected more than a thousand jobs in Norfolk.[16]

High water crept into Norfolk's industrial south side on Tuesday, June 15, and closed several businesses. As water encroached on the Vulcraft plant climbing about six inches an hour, plant manager Doyle Hopper asked for electrical service to be cut off to avoid the risk of electric shock. Other plants in the area sent workers home or were evacuated on Tuesday afternoon. Gary Pichler, manager of Island Supply, had been kept away from his business by emergency officials starting on Monday. Love Signs, a sign design and installation business, suffered damage estimated at $75,000 to $100,000 from about a foot of water than found its way into the shop. The business lost eight new computers and had to replace all its carpet and half its drywall. Love Signs had the distinction of being US Senator Ben Nelson's first stop when he visited Norfolk to assess flood damage and the possibilities for bringing federal assistance to the recovery effort. Arkfeld Manufacturing and Distributing took on about three feet of water, which damaged equipment, computers, and paper records. The waters also mixed with paint thinners, lubricants, and other toxic fluids, which not only damaged the business and made clean-up and recovery difficult, but also carried the toxic chemicals away with the receding water.[17]

The flood damaged transportation and other infrastructure across the basin. In addition to the Nebraska Central bridge, the Elkhorn knocked out others, including three former railroad bridges that had been converted for recreational use on the Cowboy Trail. The 321-mile former Chicago Northwestern Railroad right-of-way had been given to the state in 1993 for a rail-to-trail conversion, the longest in the nation. The loss of bridges and washed out sections of trail led officials to close seventy-two miles of the trail from Norfolk to O'Neill. Stanton lost the water pipe that carried drinking water from its well field, under the Elkhorn, and to the town. Replacing the pipe would cost at least $500,000 and take several months. Downriver at West Point, plumber Jerry Hugo saw increased demand for pumps as people sought to remove water from flooded basements. But other West Point business owners suffered as water covered Nebraska Highway 32 and US 275, which kept customers from reaching town.[18]

Even before the floodwaters moved into the more spacious channel downstream from Norfolk, basin residents began organizing local relief efforts. This, too, had become a regular feature of living with the Elkhorn. Norfolk residents, who had experienced their share of problems from the North Fork and the Elkhorn, formed a flood task force with the Norfolk Area Chamber of Commerce leading the charge. More than sixty representatives from local businesses and organizations met on the afternoon of June 16 to coordinate resources. In addition to the chamber's efforts, churches and charitable organizations coordinated volunteers willing to help with flood relief and solicited donations of cleaning supplies, especially liquid bleach for disinfecting flooded structures, and demolition tools for removing and replacing wet sheetrock in flooded buildings. To assist those displaced from nearly seventy area homes, the Red Cross set up comfort centers, offering water, meals, and showers, at the Norfolk City Auditorium and Elkhorn Middle School. Brewing company Anheuser-Busch donated 2,156 cases of drinking water for residents affected by the Elkhorn flood.[19] Once again, high water in the Elkhorn brought out the humanitarian character of its people.

As the record flood receded, basin residents faced unprecedented recovery. The loss of railroad service brought increased reliance on trucks, which brought more traffic to roads. Railroads not only had to replace washed-out bridges, but also repair washed out track *and* worry about track threatened by the collapse of river banks eroded by high water. To stabilize the banks, the Nebraska Central, in cooperation with the USACE and the city of Norfolk, placed the equivalent of fifty to one hundred train cars of riprap along the Elkhorn's banks. Elsewhere in south Norfolk, bank erosion threatened to dump two houses into the Elkhorn. Throughout the basin, the flood's toll mounted as the water receded and officials took stock. Holt County reported thirty bridges destroyed and miles of roads and culverts washed out. The chairman of the county board of supervisors estimated damage at $2 million. An Antelope County supervisor offered a less precise, but more dramatic report of a path of destruction a mile wide where "the roads are just gone." More than a month after the flood, county roads near the river north of Battle Creek and Meadow Grove in Madison County remained closed.[20]

At West Point, a dike protected much of the city, but spread floodwaters onto Highway 32 and farmland west of the city. Much cropland in Cuming County suffered from erosion, siltation, the drowning of crops, or some combination of those problems. A fresh topping of gravel that had been applied to many county roads in the spring had been washed away and the underlying roadbed damaged. Some culverts had to be reset and bridges at Beemer and Wisner needed repairs. The Elkhorn's strong flood current caused erosion and damage as did the uprooted cottonwood trees it carried along and rammed into structures such as bridges. West Point's Neligh Park, site of the town baseball field and the Fourth of July celebration, sat under water for more than ten days. The flooding prohibited recreation for at least those ten days and threatened to curtail celebration of the fast-approaching national holiday.[21]

Near the Elkhorn's mouth, floodwaters severely damaged flood control and bank stabilization projects along an eight-mile stretch in Dodge County. "The river went right over them," John Miyoshi, general manager of the Lower Platte North Natural Resources District, told a newspaper reporter. The structures sat underwater for about a week and a half. The flood also damaged crops and homes along a twenty-five-mile stretch in Douglas County, especially in the King Lake subdivision in the western part of the county. The Red Cross sent a dozen volunteers to help the King Lake residents cope with flood damage. Virginia Adams, a Red Cross volunteer from Kansas City, Kansas, spent a week handing out hygiene kits, cleaning supplies, and serving meals. Adams found the work to be rewarding. "There's nothing like it, to be able to help someone that really needs help," she said. "People are so grateful overall, and you can make them happy with so little." But even seven days of aid wasn't enough. Many King Lake residents still had plenty of work to do after the Red Cross ended its support and some were still displaced.[22] The relief work reflected admirable qualities in Great Plains society, but the King Lake damage also reflected the ongoing legacy of occupying the floodplain long after reliance upon water power made it an attractive location.

Late summer revealed more details about the flood's costs and a mix of old and new attitudes about living with the Elkhorn. Repairing the damaged section of the Cowboy Trail, officials now learned, would cost between $4.3 million and $6.2 million. The trail had lost a significant bridge near Norfolk and required resurfacing and debris removal over most of the sixty-seven miles between Norfolk and O'Neill. Near Clearwater, the flood-stage Elkhorn had created a new main channel, which crossed a portion of the trail it once ran alongside. Re-establishing the trail there would require rerouting two miles of trail, which meant buying property from several nearby landowners at a cost of about $1 million. The Elkhorn's wandering nature defied stable coexistence in the floodplain.[23]

The Elkhorn also jumped its banks downstream near Scribner in Dodge County and threatened to carve a new channel, to the consternation of local government officials. Steve Swanson, Scribner's mayor, had been watching the river nudging its way toward a new channel for two years before the 2010 flood. During the flood, the Elkhorn stopped nudging and leaped a hundred yards in a new direction as it washed out trees and banks. "The first part of it is probably forty acres that's all sand and silt and everything," Swanson told a reporter. "It used to be a farm field and it looks like a river channel, although the river isn't actually running there now because it's gone back down within its banks." The river didn't just remake its channel, it cut across a county road and threatened to wash out a county bridge. The road—closed for twelve days during the flood—and the bridge provided important farm-to-market access for area farmers and transportation links for schools and businesses. To stop the river's migration and to protect the transportation infrastructure, Scribner, Dodge County, the Lower Elkhorn Natural Resources District, and the USACE joined forces to begin a bank stabilization project. This interaction with the Elkhorn showed that some old ways of living with the river hadn't changed. A headline ("Officials prepare plan to fix Elkhorn River Channel issue") that accompanied newspaper reporting of the issue add to that impression. Assuming the writer used "issue" as a synonym for "problem," it suggested a perceived deficiency or fault with the river that humans would "fix" as they sought … still … to perfect their environment.[24]

At about the same time and just downstream, at the King Lake subdivision located between the Elkhorn and Platte in Douglas County, evidence emerged of a new approach to "living" with the river. In this case, though, that actually meant *not* living with the river and recognizing that maintaining structures erected in the floodplain was a losing proposition. When the flood hit King Lake on the night of June 13, the Elkhorn backed up into Rawhide Creek; both overflowed. The Douglas County Sheriff's Office called for a voluntary evacuation and used airboats to take many King Lake residents to dry land. By September, 41 of the 168 homes in the subdivision had been declared unsafe and roads throughout had been regraded and resurfaced with gravel at a cost of $21,000. Some homeowners were able to use insurance settlements to repair damaged homes. Elizabeth Townsend was among them. After three months of clean-up and repairs that cost about $35,000, Townsend was within days of moving back into her house in late September. Some of her neighbors, though, wouldn't be coming back. The Papio-Missouri River Natural Resources District offered to buy some of the unsafe homes. Those without insurance or their own funds to make repairs, might find the offer attractive or otherwise too good to refuse. Those homes, which the NRD would buy with state and, ideally federal funds, would be torn down, signaling a retreat, some might say a prudent one, from an ill-advised, costly effort to occupy the floodplain.[25]

In Norfolk, some wounds inflicted by the 2010 flood healed quickly, but others lingered for years. To the astonishment of nearly all involved, the Nebraska Central Railroad bridge re-opened fifty-eight days after it fell into the Elkhorn. Theisen Construction, Inc., a local company, began working on the bridge almost immediately. Railroaders from several states came to assist the effort aided by an over-size crane shipped in from Wisconsin and a heavy-duty bulldozer brought from St. Louis. Pulling the span from the river presented a massive challenge, but it paid off when workers discovered it had not been badly damaged. It required straightening, but that beat having to build a new structure, which would have delayed the reopening. The piers, which support the span, were another matter. The original piers, which had been sunk about 25 feet deep, had been scoured by the powerful flood current and damaged by a large tree the Elkhorn had slammed into them. Workers sank replacement piers 80 feet deep, as deep as possible, to make a considerably stronger replacement bridge.[26]

Nucor Steel, Norfolk Iron & Metal, and Louis Dreyfus Commodities, the three Norfolk manufacturers that relied most heavily on rail service (by 2010 the only line serving Norfolk was the Nebraska Central), sponsored an appreciation lunch for about 420 people to celebrate the re-opening and to thank those who had hustled to restore the bridge. Whatever the tab generated by the lunch, the companies, which combined had lost about $60,000 a day while they relied on trucks for transportation, likely saw it as a comparative pittance. Over the course of the bridge's outage, Nucor lost about $3.5 million and Norfolk Iron & Metal about $500,000.[27]

The railroad bridge's collapse left deeper wounds in the lives of Jeff Scholl's family and the emergency workers who recovered his body. The bridge went into the water on a Tuesday and, as it turned out, took Scholl with it. After more than two days of futile searching, a rescue worker spotted the body—trapped in the wrecked bridge—late in the day on Thursday. With the river still high and daylight fading, recovery workers made the difficult decision to wait for daylight to retrieve the body. The team brought in the Fire and Swift Water Rescue team from Lincoln for advice and assistance. Workers spent Friday morning removing debris and building a support scaffold. In the afternoon they brought Scholl's body to the bank.[28]

Scholl's surviving family members went through a turbulent year after his death and will likely feel its effect as long as they live. The family's story took an unexpected turn a week after Scholl's funeral when his wife, Suzie, learned she was pregnant. On February 15, 2011, Suzie delivered a girl and gave her a feminized version of her late father's name. Suzie Scholl told a newspaper reporter that news of the pregnancy brought joy and happiness back into her life and that she considered it a gift from both God and Jeff. In May 2011, Suzie Scholl filed a wrongful death lawsuit against the railroads responsible for the bridge. The suit argued that the Union Pacific Railroad, which owned the bridge and leased it, the Nebraska Central Railroad and its parent company, Rio Grande Pacific Corp., failed to ensure the bridge was designed and maintained properly. Or, one might say, the accident was an unnatural disaster. The companies answered that Scholl's death resulted from his negligence and an extraordinary force of nature—a natural disaster. The suit settled out of court two years later and terms were not reported.[29] The survivors, the Scholl family and the corporations, moved on. The Elkhorn, too, rolled on.

Flooding, the Great Plains, and the Anthropocene Epoch

If we step back one year before the ill-fated tubing expedition that opened this chapter and travel from the Elkhorn's mouth far up the basin of its next-of-kin, the Platte River, to western Nebraska, eastern Colorado, and even to the foothills of the Rocky Mountains, we can find evidence that the Elkhorn's flood history echoes across the region. Late in the summer of 2013, along the front range of the Rockies a cold front headed south from Canada collided with a moisture-laden monsoon flow headed north from the Pacific Ocean and Mexico. Beginning on September 11, 2013, the collision produced storms that dumped rain on Colorado's high country and plains. Much of the water flowing downhill found its way into the South Platte River Basin and headed east onto the Great Plains. Many streams overflowed their banks. The South Platte followed suit and blazed a trail of flood destruction across eastern Colorado and into Nebraska damaging bridges and setting records for stream flow. Even as the flood petered out in central Nebraska, the flow of contaminated water within the Platte's banks would raise concerns as far east as Grand Island.[30] That the high water did "damage" tells us that people were living with the Platte, too, in ways that might be considered ill advised. That the Platte's waters were "contaminated" tells us about perspective as well as what some of those people living with the Platte were doing.

The high water caused the greatest loss of life and property damage as it passed through areas of highest population in Colorado, but it revealed familiar issues there and even as it moved into the sparsely populated Plains.

The flood affected transportation, for example, closing portions of interstate highways, US routes, and state highways.[31] Floodwater not only passed into the South Platte Basin, but also simultaneously passed through the Denver Basin (sometimes called the Julesburg Basin), a geologic formation that runs from the front range of the Rockies onto the Plains, through which part of the South Platte flows. That in itself might not be a big deal, but the Denver Basin was home to an active petroleum industry. The flood closed nearly 1,900 oil and gas wells. And floodwaters carried petroleum and other chemicals related to its extraction from the wells into streams and rivers.[32]

The flood also turned up evidence that on Colorado's Great Plains low-income housing, and by connection low-income people, experienced special risks of suffering flood damage. In eastern Colorado the 2013 flood hit immigrants living in the United States both legally and illegally especially hard. Many of these immigrants provided critical labor for agriculture, construction, meatpacking, and service industries in the region. Many fled their mobile homes on short notice chased by rapidly rising floodwaters. They returned home to find substantial damage to their homes and property. Many, because of their immigration status, did not have flood insurance and did not qualify for federal disaster relief.

Local political and business leaders recognized the importance of these workers to the area's economy. "They're out of their homes, and a lot of them have lost their vehicles in this, they're impacted," said Lyle Achziger, mayor of Evans, Colorado. "So the economy is going to be impacted." The JBS meat-packing plant in Greeley, which employed nearly fifty immigrant workers whose families evacuated their homes to escape the flood, paid for hotels, delivered food, and otherwise tried to help those families recover. "They're certainly a critical element of our team," said JBS spokesman Cameron Bruett.[33] These stories sound familiar themes about living with a river.

On the other hand, the deluge of 2013 benefited those living with the river in the form of drought relief. Even as floodwaters carried away barns, silos, and other agricultural structures, they promised to restore soil moisture, refill reservoirs, and even to recharge depleted aquifers. "There is a silver lining if we look down the road," said Ron Carleton, Colorado's deputy commissioner of agriculture. "We just have to get past these near-term impacts."[34] Downstream in Nebraska, the debris, sewage, chemicals and other contaminants carried in the floodwater didn't seem, in the long run, to matter as much as the precious *water* that carried them. Along the Platte River agencies such as the Central Nebraska Public Power and Irrigation District and the Platte River Recover Implementation Program diverted floodwater to fill lagoons and wetlands that had been left parched by years of drought. In other places, Nebraska's Department of Natural Resources diverted water from the Platte to recharge aquifers depleted by drought and irrigation withdrawals. Floodwater also scoured the Platte's channel, which had become overgrown with shrubs, even trees, during the drought. The natural scouring provided a type of flood control, as it removed obstructions from the channel, and it deposited silt and shaped sand bars, which provided habitat for waterfowl. Dams, canals, and other water diversion projects imposed by humans living with the river had made it difficult for the river to accomplish this work on its own. As Bill Taddicken, director of the Audubon Society's Rowe Sanctuary along the Platte near Gibbon, Nebraska, put it, "The Platte River has been starved for sediment for a long time."[35] So, the high water represented a destructive calamity to some while at the same time it offered a kind of beneficial restoration to others. The river did what it does with a lot of water. Whether that was good or bad, a natural disaster, an unnatural disaster, or no disaster at all depended on perspective.

Whatever good the flood might have done for the riverbed and depleted soils and reservoirs, humans in the projected flood zone in Nebraska readied themselves for high water. For one thing, they worried and prepared. Some worried about the beneficial effects of streambed scouring, which also had a downside. The very debris the flood would scour from the riverbed would, in all likelihood, end up stacked against bridges spanning the South Platte. As we've seen time and again with the Elkhorn, the debris would turn the bridges into dams that could, in turn, send the high streamflow out of the banks and onto surrounding agricultural, residential, or even urban land. "Debris in the water is going to be the biggest issue and could contribute to additional flooding," said Brian Dunnigan, then-director of the Nebraska Department of Natural Resources.[36] Natural river action, deemed by some restorative, stood to be converted into destructive unnatural disaster.

Others worried about how the landscape had changed in western Nebraska since the last major Platte flooding in that region, which happened in the mid-1960s. Since then, Interstate 80 had been completed. Parts of the roadbed sat in the floodplain and several times it crossed the Platte. No one knew how the water would behave in those places where the road and the river coexisted in the floodplain.[37]

Still others channeled their concern and planning into action. In the places most likely to face floodwaters, communities built emergency dikes, placed sandbags to protect wastewater treatment plants, sewer systems, water supplies, telecommunications equipment, and other vital infrastructure projects. Those farmers who could brought in harvests early. Homeowners moved valuables to high ground. The Union Pacific Railroad elevated seven miles of track near Ogallala, Nebraska.[38] The railroad wanted to keep its traffic moving, which meant keeping that flood off its tracks.

Worrying and preparation paid off at Hershey, Nebraska. There the Platte broke through its banks and flowed into the southeast part of the town. But a flood berm and emergency sandbagging helped keep floodwaters out of homes. A local official credited state assistance, in the form of early flood warnings and sandbags, for averting costly flood damage. Similar stories unfolded in Big Springs, Brule, and Ogallala.[39]

The Platte registered flood stage at Kearney, but didn't do much damage there or downriver; nevertheless, the flood raised concerns as far east as Columbus. There officials warned that it could be days or even weeks before it would be safe to enter the Platte River again. State testing showed bacteria levels in the Platte in western Nebraska reached ten to fifteen times the amount considered "high risk," meaning a person would be more likely to become ill after consuming the water. In addition to bacteria, officials worried that the floodwaters could contain petroleum, agrichemicals, animal waste, untreated human waste, and even dissolved metals from mining, likely from far upstream in Colorado.[40] All those ingredients carried in the Platte had the potential to turn the river into a calamity even if it didn't leave its banks. Would that have qualified as a natural disaster or an unnatural one?

As the flood of 2013 left western Nebraska and headed downhill for the Missouri River and points beyond, one Plainswoman, an unnamed staff writer for the *North Platte Telegraph* and the wife of a local farmer, reflected on the flood's complicated meanings. Immediately she recognized "the great power and devastation" evident in the flowing water. But quickly she pivoted to curiosity inspired by the freak event. Onlookers and gawkers, gathered on bridges in western Nebraska were drawn to behold the awesome power and size of a familiar thing swollen to unseen and perhaps unimagined proportions. The writer knew well the flood's record of death and destruction, yet she and others couldn't "get enough of looking at what the flow of the river looks like and the power that it has over the land."[41] That made her feel guilty.

She and her husband had dealt firsthand with the previous big Platte flood of recent memory, which happened in 1995. They "survived" that flood, as she put it, and were lucky not to have lost more than a hay field. "But from that flood we learned how to be more prepared for any other big floods that may happen again," she wrote. And so they were. On a Saturday in 2013 they made their own river gauge so they could watch and measure the river's rise. They could see the first wave of floodwater arrive because it carried "brown muck" that contrasted with the normally clear Platte. By the following Tuesday, the river was falling and they decided to tour their flooded land in kayaks, to go on a flood adventure. They paddled down their driveway, over their hay field, and down the county road, floating across the Great Plains on portions normally flooded by a sea of grass.[42] Too much water had turned the sub-humid, treeless world of the Great Plains upside down.

The Platte flood of 2013 was one of the worst in decades and by many accounts the Elkhorn flood of 2010 was the worst one ever seen in parts of that basin. What's going on? Have Plainsfolk been living through the most recent incarnation of the volatile mid-continent climate pattern for which the Great Plains is famous? Is the climate becoming less sub-humid? Is a new climate regime taking hold?

Donald Worster, an important interpreter of American water history and the environmental history of the Great Plains, has addressed these questions. In the early twenty-first century, Worster trained his formidable intellect on what's happening to the West, specifically what climate change might do to the American West. His meditation on the possibilities took him right to a word at the core of this project: disaster. The word disaster (great destruction, hardship, loss of life) comes to us from fifteenth-century Italy, which joined the Latin roots *dis* and *astro* to describe the concept of a "bad star." People of that time and place considered certain stars omens of bad things from drought to pestilence to flood. They would have considered all such things natural manifestations of an angry God punishing his sinful children. Things evolved during the Enlightenment as people began to accept more secular explanations for calamities. Floods, earthquakes, and the like came to be seen as the result of natural forces beyond human control and indifferent to humans. With humans assuming the role of innocent victims of an indifferent nature, anything humans might do to control or contain those forces made sense. Not everyone made the transition. In 2011 Rick Perry, governor of Texas at the time, proclaimed three days of prayer asking God to send rain to his drought-ravaged state, a considerable portion of which rests in the Plains. Governor Perry notwithstanding, human understanding of the species' relationship to nature continues to evolve with some thinkers having reached a postmodern paradigm. Postmodern society doesn't separate humans from nature; "the whole world is now humanized," as Worster put it.[43] That might not be a bad thing, as humans are animals whose habitat is Earth. But, where does that leave us in trying to make sense of the relationship between people and, say, rivers?

One way to consider it would be to place the relationship in a new age. Scientists Paul J. Crutzen and Eugene F. Stoermer suggest we call the period in which we live the *Anthropocene*. They intend it to describe the latest epoch in Earth's history "when humans are the definitive shaping force and nature is no longer an autonomous Other in our lives."[44] Those who use the term typically date the era to the Industrial Revolution and the heavy use of fossil fuels. Others peg it to July 16, 1945, when scientists detonated the first nuclear bomb. The new epoch brings with it a new sense of disaster that includes massive pollution of the oceans and skies and the threat of mass extinctions of plants and animals, all of which can be laid squarely at the feet of humans.[45] Maybe living with rivers on the Great Plains should be cast in the Anthropocene, too. How else to consider petroleum and pollution from mine tailings produced in Colorado taking a ride in the Platte on their way to posing a health risk to people in Columbus, Nebraska?

All this might seem far removed from the Elkhorn River Basin, which resides at the end of the line of the Mississippi River's network of tributaries and nestled in the northeast corner of Nebraska. Or maybe not. Thought of another way, humans have been pretty seriously hammering out the Anthropocene in the Elkhorn Basin for decades even if they haven't quite achieved it. Maybe the Elkhorn sits at the end of the line in these terms, too, where humans aren't quite the definitive shaping force and where nature still exerts a fair measure of autonomy. Thought of in still another way, as part of the global hydrologic cycle, the Elkhorn actually sits somewhere on a circle. Surface water evaporates, goes into the atmosphere, falls to earth as rain, perhaps drains into a stream or river if that rain falls on land, flows into lake, sea, or ocean, and repeats the journey. Considered in those terms, the Elkhorn and its neighborhood of the Great Plains are part of a much bigger story. Should the human-made disaster of global warming come to the Plains, the low spot in the terrain through which the Elkhorn flows will be part of the narrative.

On the heels of having triumphed over nature, according to the postmodern narrative, humans will have to deal with human-made disaster born of what Worster calls "our high-energy way of life."[46]

That way of life might be ushering in a new climate regime born of global warming and might feature melting glaciers, less snow, more rain, and more rivers overflowing their banks. Already, Worster notes, "we are flooding large areas of the planet."[47] Turning our attention briefly back to Hannibal, Missouri, we find the first recorded major flood there happened in 1851. The recent past, though, has seen the problem of flooding grow worse. Of the city's ten worst floods recorded through 2000, *nine* have happened since 1960.[48] What will that mean for the Great Plains? It's hard to say. The central continent weather pattern was considered extreme *before* global warming was identified. Worster predicts a future Nebraska with an average summer temperature of 85 degrees, compared to the current 72 degrees, and the hottest days of summer exceeding 120 degrees. The state would still freeze in winter, but not drop far below zero as in the past. For the American West generally speaking, the biggest problem won't be heat, but drought. Optimistic scenarios predict a more desert-like West. The *pessimistic* ones envision droughts lasting for decades, if not centuries, and a massive exodus to points cooler and moister.[49]

It might seem counterintuitive, but drought can make for nasty flooding. In January 2001, natural resource managers in Nebraska augured through river ice on the Platte in an effort to divine what that year's flood season might hold. They found ice ranging from twelve to eighteen inches thick and averaging about fifteen inches. That first sign wasn't good. What they found beneath the ice—little water flowing in the river—might have been worse. They found less water than they had in the bad flood years of 1993 and 1997. In 1997 state officials resorted to dynamite to break up a mile-long ice jam that threatened to cut off Lincoln's drinking water supply. The low flow in 2001 reflected drought conditions—and raised concerns about flooding. Lots of ice and little water to push it downstream could lead to massive ice jams and lowland flooding in the spring thaw. "It doesn't mean [flooding's] going to happen," explained Steve Oltmans, manager of the Papio-Missouri River Natural Resources District. "But the odds are we are going to have some problems. It's just a matter of time."[50]

The spring of 2001 didn't leave an impression in the state's flood record book, but this little bit of information tells us some important things about the state of living with rivers on the Plains. First, winter meetings to discuss how to handle ice on the rivers had become a regular part of flood preparation and state and local agencies stood ready to try to manage river flow and to prevent flooding. More importantly, this reveals what could be an unexpected consequence of climate change. If the predictions cited by Worster are correct, scorching heat and increased aridity might not be the only problems facing Plainsfolk in the future. Some variation on the thick-ice-pack-and-drought-caused flood scenario could also be part of an intensified climate experience. But, it's not the job of historians to predict the future, so let's consider where the past has left us.

By envisioning the interaction between humans and the Elkhorn Basin's riparian environment as a dialogue, one can imagine a conversation that began about 10,000 years ago. For the overwhelming majority of that time, humans listened to the environment, so to speak. They had little choice. During these years, Native people hunted and farmed the basin's floodplains, but usually lived on terraces that were beyond danger of flooding and which were easier to defend than bottomlands. As the basin's residents built progressively more substantial dwellings, they were increasingly likely to locate them above the floodplain, presumably to protect their valuables from damage.

The arrival of non-Indian settlers in the mid-nineteenth century marked a change in the dialogue's nature. At first, the basin's latest arrivals could only react to the river's floods. Travelers such as the Duke of Württemberg and founders of the basin's first non-Indian towns altered their routes when the Elkhorn overflowed. Likewise, these recently arrived farmers could do little other than clean up after an inundation and hope that their bottomlands would dry early enough to sustain a replanting.

In addition to fertile farmland, the basin offered its newest inhabitants easy overland transportation and hydraulic power for mills. Humans responded by traveling up the valley on wagon roads and railroads. They harnessed the river's power to operate mills, pump water, and generate electricity. Commerce developed, which lured more settlers and spawned more towns, which, in turn, led to further intensification in using the valley's resources. The agricultural techniques of the recently arrived culture, for instance, led farmers to supplant native grasses and Indian bottomland farming with row crops in huge fields extending into the higher elevations. This disturbed the basin's natural hydraulic cycle and its built-in checks on runoff. The extent to which this influenced flooding in the basin is difficult to determine, but residents who had spent much of their lives in the basin testified that, in their minds, flooding grew worse during their lifetimes. Unfortunately, lack of records before non-Indian settlement makes true long-range comparisons impossible. Furthermore, considered against the backdrop of geologic time, these impressionistic reminiscences have limited value. At this phase, one might say the nature of flood damage changed. Members of the dominant society wouldn't have seen the river as angry, but they might have considered a flood a sign of retribution or punishment for their sins according to the new cultural regime they brought with them. We might consider this both an era of natural disaster, probably the dominant explanation for flood, and the dawn of the era of unnatural disaster, as people consciously placed assets in the floodplain and set out simultaneously to defend their newly claimed territory and to reform or perfect a natural entity, the Elkhorn, that almost overnight had come to be viewed as not good enough.

The most significant development was that the newly arrived culture brought the technology to alter the river and the propensity to do it. Within a generation or two of establishing farms, mills, towns, and transportation networks on the floodplain, humans undertook to protect these assets from flooding. Initially, the new culture assumed a localized, defensive posture in relation to the river and its tributaries—by 1888 they had built levees along some of the banks to divert floods. Shortly after the turn of the century, humans took the offensive in their relationship to the Elkhorn, but continued to approach flood control in a piecemeal fashion. Local drainage districts dug ditches or "improved" existing stream and river channels by straightening or deepening them to facilitate the flow of water from their jurisdiction. Drainage districts scored some local successes, but tended merely to divert water downstream where it caused more damage than was typical prior to the districts' actions. Costly litigation from such damages drove a number of districts out of business by the 1930s as drought and the Dust Bowl helped to place flooding far from the minds of most people on the Great Plains. Very quickly during this phase, people went from depending on the Elkhorn for power and ice and recreation to having little practical need for the river in their daily lives.

When basin residents and local government agencies returned their attention to floods, they did so in a more unified fashion and in partnership with state and federal governments. The New Deal atmosphere of the 1930s was especially important as it spawned an unprecedented expansion of federal power, including flood control projects. The US government extended its jurisdiction from only navigable streams to include streams that, when flooded, posed a threat to transportation and interstate commerce. The record-breaking flood of 1944 brought the federal government's expanded power to bear on the Elkhorn Basin.

Responding to the House of Representatives, the USACE investigated 1944's flood damages basinwide. Soliciting testimony from residents and local governments, the Corps produced the most thoroughly researched and comprehensive flood control recommendations for the Elkhorn Basin to date. Although the Corps appeared to favor developing the basin with an eye toward flood control supplemented by irrigation and power production, the cost of such projects and the ability to garner local support for them led the Corps to recommend local protection projects. These projects relied mostly on levees to protect cities and towns, but they also included channel modifications that were relatively minor except for the Norfolk project. In spite of the apparent benefits of a basinwide response to flooding, the 1940s saw humans revert to a mostly defensive, localized approach to changing the river, but in partnership with the federal government and its immense resources.

This left much of the Elkhorn's floodplain open to continued development and that development vulnerable to flooding. The same culture that brought these people into the floodplain for economic benefit kept them there and saw them protect their property and interests as dictated by economics. The 1960s saw a boom in the construction of flood protection and mitigation projects, a number of which had been proposed by the Corps in the 1940s but had languished. Nagging flood problems finally snapped resistance to the projects, especially ones in or near the most densely populated areas of the basin. New federal (National Flood Insurance Program, for example) and state (creation of natural resource districts) approaches contributed to changes from the late 1960s and early 1970s through the date of this writing. These changes cropped up at about the same time as the environmental movement and a growing awareness of environmental problems associated with the Anthropocene. In terms of flood control it finally dawned on people that old approaches had serious drawbacks and that it was time to go in a new direction. As Christine Klein and Sandra Zellmer write, "We have wasted more than a century pursuing a foolish ideal: *the floodless floodplain*." They continue, "We struggled to hold back the water from the people, . . . [but] gave insufficient attention to the opposite notion—holding back the people from the waters."[51]

Alan Lightman, a physicist who teaches humanities at the Massachusetts

Institute of Technology, reflecting on tornadoes and mudslides that inflicted devastating damage in the South, Midwest, and Pacific Northwest in 2014, concluded that those events, to which floods could certainly be added, left humans feeling betrayed by nature. Lightman asked, "How could Mother Nature do this to us, her children?" Whatever connection we feel to nature and whatever we think about nature, nature doesn't care a bit about us. Calamities such as mudslides, tornadoes, and floods "happen without the slightest consideration for human inhabitants." Nature is neither our friend nor our enemy. "Nature simply is," Lightman reminds us. However we might feel about nature and whatever personality traits we might ascribe to it, "[t]here is no mind on the other side of the wall."

With respect to a report by the United Nations Intergovernmental Panel on Climate Change, which documents damages caused by human-created greenhouse gases and global warming, he says we shouldn't be worried about protecting our planet. Whatever we do, it will go on and it doesn't care what happens to us. Rather, we should be worried about protecting ourselves "because we have only ourselves to protect us."[52]

So what does this all mean? Reviewing well more than a century's record of human interactions with the Elkhorn and its tributaries suggests that there really is no mind on the other side of "the wall." The Elkhorn didn't care about washing out roads or railroad tracks or mill dams. It didn't think about collapsing the railroad bridge that killed Jeff Scholl. It makes all the sense in the world that we should protect ourselves because we can and we have only us to do it. In that light, the policy changes and approaches to living with the Elkhorn that have emerged in recent decades show glimmers of light in an all-too-often dark story. As Klein and Zellmer put it, "It's time to try something different: *giving rivers room to flood*."[53] To put it another way, getting people and the trappings of civilization out of the floodplain would go a long way toward ushering out the era of the unnatural disaster.

Can it be *that* difficult to get people out of the floodplain? Some parts of the basin's infrastructure can't in any practical way be removed entirely from the floodplain. But, people can put more distance than they have between themselves and their property and the Elkhorn's highest reach. As much sense as that might make, it would put people in the position of acting against two powerful propositions: the element in our culture that says we can conquer nature and the notion that the legacy of the indomitable hearty pioneers who heroically conquered the Plains cannot be repudiated. So, yes, it is difficult to get out of the floodplain. If the predictions about climate change are on target, the floodplain might be expanding, which would make getting out of the river's way even more difficult. The relationship between people and the Elkhorn seems destined to continue with people doing the best they can. Perhaps they can ask themselves how they wish to write the next chapter in the story. One version could echo the sense of resignation found in the last stanza of Bob Dylan's "Buckets of Rain": "Life is sad/Life is a bust/All ya can do is do what you must." That would be disheartening. But the song doesn't end there. Like the river and its people, it rolls on and points to a more uplifting possibility for the relationship: "You do what you must do and you do it well/I do it for you/ Honey baby, can't you tell?"[54]

NOTES

Introduction

1. Bob Dylan, "Buckets of Rain," *Blood on the Tracks*, LP, Columbia PC 33235, 1975; William Ruhlmann, review of "Tangled Up in Blue," *Allmusic*, http://allmusic.com/song/tangled-up-in-blue-mt0004719211 (accessed 1 June 2015).
2. See Todd M. Kerstetter, "Up the River: Flooding and Flood Control in Nebraska's Elkhorn River Basin through the Floods of 1944" (master's thesis, University of Nebraska, 1992); Kerstetter, "High Water: Flooding in the Elkhorn Basin, 1823–1940," *Nebraska History* 76 (Winter 1995): 176–187; and Kerstetter, "'The Worst Floods in History': The Federal Government and the Floods of 1944 in the Elkhorn River Basin," *Great Plains Quarterly* 21 (Summer 2001): 179–192.
3. Stanley Johnson, "How We Do Things at Pierce, Nebr.," postcard, Waupun, WI, 1910, author's collection; "Night Scene on Elkhorn River, Clearwater, Nebraska," postcard, Nebraska State Historical Society, RG2986.PH000002-000007; "Ta-Ha-Zouka Park, Norfolk, NE," postcard, Chicago: Curt Teich Company, 1941 [Teich catalog number 1B378-N].
4. Emanuel Wolfe, [Kids swimming in the Elkhorn River], glass plate negative, n.d., Nebraska State Historical Society, RG2836.PH0-000296; Michael A. Amundson, "Portrait of a Small Town: The Photographic Diary of Neligh's Emanuel Wolfe," *Nebraska History* 79 (Winter 1998): 150–151.
5. Norm Bomer, *Sons of the River: A Nebraska Memoir* (Moscow, ID: Canon Press, 2000), 13.
6. Ibid., 75.
7. Karen G. Jackovich, "Carson's Homecoming," *People Weekly* 16 (Dec. 7, 1981): 71–2+, accessed 4 November 2014, http://www.people.com/people/archive/article/0,,20080870,00.html; John Miller, "From the Great Plains to L.A.: The Intersecting Paths of Lawrence Welk and Johnny Carson," *Virginia Quarterly Review* 79 (Spring 2003): 265–279.
8. I wish not to burden general readers with a historiographical discussion in the text, so I've confined much of it to the notes. Superb work on rivers outside the United States

includes Mark Ciocc, *The Rhine: An Eco-Biography, 1815–2000* (Seattle and London: University of Washington Press, 2002), and Grace Karskens, *People of the River* (Sydney: Allen & Unwin, forthcoming).

9. See Ari Kelman, *A River and Its City: The Nature of Landscape in New Orleans* (Berkeley: University of California Press, 2003), and Craig E. Colten, *An Unnatural Metropolis: Wresting New Orleans from Nature* (Baton Rouge: Louisiana State University Press, 2005). Perhaps the best recent treatment of a significant portion of a major river's basin can be found in Christopher Morris, *The Big Muddy: An Environmental History of the Mississippi and Its Peoples from Hernando de Soto to Hurricane Katrina* (New York: Oxford University Press, 2012). For an inspiring treatment of a different urban area and its relationship to a river, see Jared Orsi, *Hazardous Metropolis: Flooding an Urban Ecology in Los Angeles* (Berkeley: University of California Press, 2004).

10. Those interested in the general history of water in the trans-Mississippi West should begin with Norris Hundley, Jr., "Water and the West in Historical Imagination," *Western Historical Quarterly* 27 (Spring 1996): 5–31; Norris Hundley, Jr., "Water and the West in Historical Imagination: Part Two—A Decade Later," *The Historian* 66 (Fall 2004): 455–490; and Norris Hundley, Jr., *The Great Thirst: Californians and Water, 1770s–1990s* (Berkeley: University of California Press, 1992). Works on storage, irrigation and drought include Donald Worster, *The Dust Bowl: The Southern Plains in the 1930s*, 25th Anniversary Edition (New York: Oxford, 2004); R. Douglas Hurt, *The Dust Bowl: An Agricultural and Social History* (Chicago: Nelson-Hall, 1981); Donald J. Pisani, *From the Family Farm to Agribusiness: The Irrigation Crusade in California and the West, 1850–1931* (Berkeley: University of California Press, 1984); James Earl Sherow, *Watering the Valley: Development along the High Plains Arkansas River, 1870–1950* (Lawrence: University Press of Kansas, 1990); Donald Worster, *Rivers of Empire: Water, Aridity, and the Growth of the American West* (New York: Oxford, University Press, 1985); B. Lynn Ingram and Frances Malamud-Roam, *The West without Water: What Past Floods, Droughts, and Other Climatic Clues Tell Us about Tomorrow* (Berkeley: University of California Press, 2013); and William deBuys, *A Great Aridness: Climate Change and the Future of the American Southwest* (New York: Oxford University Press, 2011). Those interested in the relationship between a river and a more rural western landscape outside the Great Plains will find an exemplary model in Nancy Langston, *Where Land and Water Meet: A Western Landscape Transformed* (Seattle and London: University of Washington Press, 2003).

11. See Robert Kelley Schneiders, *Unruly River: Two Centuries of Change along the Missouri* (Lawrence: University Press of Kansas, 1999). In addition to works on rivers and flooding, this book contributes to regional history by providing detailed environmental history of a river basin that adds to fundamental works such as Webb's *The Great Plains*, and R. Douglas Hurt, *The Big Empty: The Great Plains in the Twentieth Century* (Tucson: University of Arizona Press, 2011). Hurt offers the most recent attempt at

synthesizing the Great Plains history and he promises "to show how the people of the Great Plains attempted to control the environment during the early twentieth century but later settled for just trying to conserve and preserve it." Whatever the book's achievements, it says little about rivers, flooding, and flood control beyond its fine discussion of the Pick-Sloan Plan for the Missouri River. See Hurt, xiv, 179–186. Two recent books offer comprehensive environmental histories of southern Plains rivers. Patrick Dearen, *Bitter Waters: The Struggles of the Pecos River* (Norman: University of Oklahoma Press, 2016) expertly chronicles the life of the Pecos, which flows through New Mexico and Texas. He carefully identifies the river's wildly different segments and examines difficulties arising from water quality (highly saline in places), efforts to tap the Pecos for irrigation, degradation of the river and riparian environment, and issues related to flow, from low and resulting problems with storage and irrigation, to high and the destructive force of floods. Ultimately, Dearen's book points to possibilities for the Pecos's future. Can humans use it to serve various purposes without effectively killing it? Can it be saved and restored? If so, how? Restoration to standards of pre-European settlement would require coordinated holistic treatment of the entire basin, which seems unlikely. As with many Plains rivers, precious water flowing through arid land beckons use and tempts people to push use to unwise and destructive lengths. For treatment of the Brazos River, see Kenna Lang Archer, *Unruly Waters: A Social and Environmental History of the Brazos River* (Albuquerque: University of New Mexico Press, 2015).

12. Walter Prescott Webb, *The Great Plains* (Boston: Ginn, 1931), 3, 17.
13. The Plains climate swings from periods of drought that can last several years to periods of above-average rainfall. See Andrew C. Isenberg, *The Destruction of the Bison: An Environmental History, 1750–1920* (Cambridge: Cambridge University Press, 2000), 17.
14. Dan Haas, "Wild and Scenic Rivers," in *Encyclopedia of the Great Plains*, David J. Wishart, ed. (Lincoln: University of Nebraska Press, 2004), 866–867; "Water" in *Encyclopedia of the Great Plains*, 844–867; Paul R. Jordan, "Rivers," in *Encyclopedia of the Great Plains*, 861–862.
15. Elliott West, *The Contested Plains: Indians, Goldseekers, and the Rush to Colorado* (Lawrence: University Press of Kansas, 1998), 36. West provides a remarkable discussion of the ecological importance of river valleys throughout this book, but especially on pages 207–235. See also Elliott West, *The Way to the West: Essays on the Central Plains* (Albuquerque: University of New Mexico Press, 1995), 13–50.
16. People disagree about the boundary of the Great Plains. The Elkhorn's headwaters rest comfortably within the unambiguous Plains, but its mouth sits at or even beyond what some consider the region's eastern boundary. Using the definition provided by the *Encyclopedia of the Great Plains*, which places the region's eastern boundary at the eastern border of the state of Nebraska, the Elkhorn flows entirely within the Great Plains.

David J. Wishart, "The Great Plains Region," in David J. Wishart, ed., *Encyclopedia of the Great Plains* (Lincoln: University of Nebraska Press, 2004), viii–xviii.

17. Ted Steinberg, *Acts of God: The Unnatural History of Natural Disaster in America* (New York: Oxford University Press, 2000), xv–xvii.
18. Ibid., xvii.
19. Ibid., xix, xxiii.
20. Christine A. Klein and Sandra B. Zellmer, *Mississippi River Tragedies: A Century of Unnatural Disaster* (New York: New York University Press, 2014), 11.

Chapter 1

1.James C. Malin, "Ecology and History," in James C. Malin, *History and Ecology: Studies of the Grassland*, Robert P. Swieringa, ed. (Lincoln: University of Nebraska Press, 1984), 107.
2. Joseph Lathrop, D. D., "Moral Reflections on Floods," Sermon preached at the First Church in West Springfield, MA, 22 January 1807 (Springfield: H. Brewer, Printer, 1807).
3. Ainsworth, Nebraska, the community closest to the Elkhorn's source, bills itself as "The Middle of Nowhere."
4. Nebraska State Planning Board, *Water Resources of Nebraska* (Lincoln, 1941), 168–169.
5. Nebraska State Planning Board, 168–169. For additional information on Great Plains water and rivers, see James E. Sherow, "Water," in *Encyclopedia of the Great Plains*, 845–850, and Paul R. Jordan, "Rivers," in *Encyclopedia of the Great Plains*, 861–862.
6. Ray Bentall, *Water Supplies and the Land: The Elkhorn River Basin of Nebraska*, Conservation and Survey Division Resource Atlas No. 1 (Lincoln: University of Nebraska, 1971), 18.
7. Ann Bleed and Charles Flowerday, eds., *An Atlas of the Sandhills*, 2nd ed. (Lincoln: Conservation and Survey Division, Institute of Agriculture and Natural Resources, and University of Nebraska-Lincoln, 1990), 1.
8. Ibid., 1–15.
9. Bentall, *Water Supplies*, 17.
10. Ibid., 2–6.
11. Bob Grier, "The Elkhorn," *Nebraskaland*, spec. ed., *Nebraska Rivers* 61 (January–February 1983): 69.
12. Bentall, *Water Supplies*, 2.
13. Ibid., 14–15.
14. Ibid., 29.
15. Walter Prescott Webb, *The Great Plains* (Boston: Ginn, 1931; reprint, Lincoln: University of Nebraska Press, 1981), 17.
16. Grier, "The Elkhorn," 69.

17. Bentall, *Water Supplies*, 20.
18. Nebraska State Planning Board, *Water Resources of Nebraska* (Lincoln, 1941), 169.
19. Bentall, *Water Supplies*, 3.
20. Bleed and Flowerday, *An Atlas of the Sandhills*, 20, 127.
21. Bentall, *Water Supplies*, 4, 6.
22. Bleed and Flowerday, *An Atlas of the Sandhills*, 27–28. Atmospheric demand for moisture is referred to as the potential evapotranspiration, the combined total evaporation of freestanding water and the transpiration of moisture from the soil through vegetation. As measured at Ewing, in eastern Holt County, evapotranspiration exceeds precipitation, increasingly so, from mid-February until early July when the two measurements begin to close on each other, becoming equal in late October.
23. Grier, 69–70.
24. Bentall, *Water Supplies*, 3–6.
25. Ibid., 39.
26. Ibid., 24.
27. Ibid., 39.
28. Nebraska Natural Resources Commission, Water Quality Planning Staff, *Elkhorn River Basin Water Quality Management Plan* (Lincoln: Nebraska Natural Resources Commission, 1975), 4–3.
29. Ibid., 4–3.
30. Ibid., 4–3 through 4–6.
31. For an excellent discussion of different types of floods and how they create greatly different perceptions and reactions, see Grace Karskens, "Floods and Flood-mindedness in Early Colonial Australia," *Environmental History* 21 (April 2016): 315–342.
32. Luna B. Leopold, *Water: A Primer* (San Francisco: W. H. Freeman and Company, 1974), 72.
33. William G. Hoyt and Walter B. Langbein, *Floods* (Princeton, NJ: Princeton University Press, 1955), 12–18.
34. Leopold, *Water*, 73.
35. Hoyt and Langbein, *Floods*, 9.
36. Ibid., 18–22.
37. Ibid., 22–23.
38. Ibid., 27–31.
39. Ibid., 31–34.
40. Lloyd R. Bailey, *Noah: The Person and the Story in History and Tradition* (Columbia: University of South Carolina Press, 1989), 5–6.
41. Gen., 2:4–8:20. Hoyt and Langbein estimate the biblical flood occurred in 2957 BCE. Hoyt and Langbein, *Floods*, 3.
42. Hoyt and Langbein, *Floods*, v, 5–6.
43. Karskens, "Floods and Flood-mindedness," 319.

44. Ibid., 319–323.
45. Ibid., 331.
46. Ibid., 332–333.
47. Ibid., 337–338.
48. Battiste Good Winter Count, "Four lodges drowned winter (1711–1712)," *Lakota Winter Counts: An Online Exhibit*, Smithsonian Institution, National Anthropological Archives, and National Museum of the American Indian [2002], accessed 26 January 2016, http://wintercounts.si.edu/flashindex.html.
49. Elizabeth A. Fenn, *Encounters at the Heart of the World: A History of the Mandan People* (New York: Hill and Wang, 2014), 68, 286–287.
50. Lakota winter counts for 1825–1826 recording the flood include American Horse, "Dakotas living south of Whetstone Agency killed when the Missouri River flooded"; Battiste Good, "Many Yanktonais drowned winter"; Cloud Shield, "Many of the Dakotas were drowned in a flood caused by a rise in the Missouri River, in a bend of which they camped"; Long Soldier, "Year that ice broke causing flood and nearly half camp drowned"; No Ears, "They drowned"; and Swan, "Thirty lodges of Dakota Indians drowned by a sudden rise of the Missouri River," *Lakota Winter Counts: An Online Exhibit*, Smithsonian Institution, National Anthropological Archives, and National Museum of the American Indian [2002], accessed 26 January 2016, http://wintercounts.si.edu/flashindex.html.
51. Fenn, *Encounters at the Heart of the World*, 286–287; Battiste Good winter counts, "Camped cutting the ice through winter" (1702–1703) and "Eat frozen fish winter" (1748–1749), *Lakota Winter Counts*, http://wintercounts.si.edu/flashindex.html, accessed 26 January 2016.
52. Ray H. Mattison, "The Flood of 1881," *North Dakota History* 34 (Summer 1967): 227, 229, 231, 233–237.
53. Ibid., 231, 233, 235–239, 241.
54. Ibid., 231, 236, 240–241; John E. Carter, "Niobrara, Nebraska: The Town Too Tough to Stay Put!" *Nebraska History* 72 (Fall 1991): 145. A detailed analysis of ice and floods can be found in Herbert T. Hoover, John Rau, and Leonard Bruguier, "Gorging Ice and Flooding Rivers: Springtime Devastation in South Dakota," *South Dakota History* 17 (Summer–Fall 1987): 181–201.
55. Marian Ridgeway, *The Missouri Basin's Pick-Sloan Plan: A Case Study in Congressional Policy Determination*, Illinois Studies in Social Science, vol. 35 (Urbana: University of Illinois Press, 1955), 3–4.
56. Ibid., 7.

Chapter 2

1. Klein and Zellmer, *Mississippi River Tragedies*, 187.
2. The account does not name the Indians, who were presumably Omahas. Paul

Wilhelm, Duke of Württemberg, *Travels in North America, 1822–1824*, trans. W. Robert Nitske, ed. Savoie Lottinville (Norman: University of Oklahoma Press, 1973), 334–335.

3. Amy E. Harvey, *Oneota Culture in Northwestern Iowa* (Iowa City: University of Iowa, 1979), 36; J. Rob Bozell, Curator of Archaeology, Nebraska State Historical Society, interview with the author, Lincoln, NE, 20 January 1992. Harvey's book provided the dates during which various cultures inhabited the region, while Bozell indicated which cultures have been detected in the Elkhorn Basin.
4. Arrell Morgan Gibson, *The American Indian: Prehistory to Present* (Lexington, MA: D. C. Heath and Company, 1980), 21–25.
5. Harvey, *Oneota Culture*, 36; Bozell, interview.
6. Donald J. Blakeslee and Warren W. Caldwell, *The Nebraska Phase: An Appraisal* (Lincoln, NE: J & L Reprint Company, 1979), 20, 25; Donald J. Blakeslee, "Assessing the Central Plains Tradition in Eastern Nebraska: Content and Outcome," in Donald J. Blakeslee, ed., *The Central Plains Tradition: Internal Development and External Relationships* (Iowa City: Office of the State Archaeologist, The University of Iowa, 1978), 134.
7. John Ludwickson, "Central Plains Tradition Settlements in the Loup River Basin: The Loup River Phase," in Blakeslee, ed., 96–97.
8. Waldo R. Wedel, *Prehistoric Man on the Great Plains* (Norman: University of Oklahoma Press, 1961), 94–95.
9. Ludwickson, "Central Plains Tradition Settlements," 96–97.
10. Wedel, *Prehistoric Man*, 100–101.
11. Ibid., 117–120.
12. Alice Fletcher and Francis La Flesche, *The Omaha Tribe*, 2 vols. (Washington, DC: Government Printing Office, 1911; reprint, Lincoln: University of Nebraska Press, 1972), 1:36.
13. There is no English translation of "Waté." See Fletcher and La Flesche, 1:89.
14. Wedel, *Prehistoric Man*, 80.
15. Fletcher and La Flesche, *The Omaha Tribe*, 1:88, 89.
16. G. Hubert Smith, "Notes on Omaha Ethnohistory, 1763–1820," *Plains Anthropologist* 18 (November, 1973): 257–264; Paul Wilhelm, Duke of Württemburg, *Travels in North America, 1822–1824*, trans. W. Robert Nitske, ed. Savoie Lottinville (Norman: University of Oklahoma Press, 1973), 313. Stating who lived in the Elkhorn Valley at what time is difficult. According to Smith, Louis Joliet and Father Jacques Marquette made the first mention of the Omahas in 1673–74. They reported the Omahas east and north of the Missouri and in possible contact with the Elkhorn Valley until the late eighteenth century. The Poncas are generally associated with the Niobrara River and its valley (adjacent to the Eklhorn Valley on the north), but in 1823 the Duke of Württemburg encountered a Ponca hunting party not far above the Omaha village

on the Elkhorn River. Joseph Jablow, *Ethnohistory of the Ponca* (New York: Garland Publishing, 1974), 349.

Likewise, Pawnees are typically associated with the Platte and Republican rivers, but their hunting ranges extended north and east to include portions of the Elkhorn Valley. Richard White, *The Roots of Dependency: Subsistence, Environment, and Social Change among the Choctaws, Pawnees, and Navajos* (Lincoln: University of Nebraska Press, 1983), 150.

Homelands of Plains tribes fluctuated, and the association of Omaha, Ponca, and Pawnee with the Elkhorn Valley reflect homelands at the time of white contact in the eighteenth and nineteenth centuries. In 1859, Ponca hunters reportedly encountered a combined party of Brulé, Oglala, and Cheyenne at the Elkhorn's headwaters. Bleed and Flowerday, eds., *An Atlas of the Sandhills*, 196.

17. Paul Wilhelm, Duke of Württemburg, 333.
18. Fletcher and La Flesche, *The Omaha Tribe*, 2:623. This source is unclear about Omaha residence on the Elkhorn. In volume two, the authors write that by 1830 Sioux attacks had driven the Omahas from their Missouri River villages to the southwest, where they established villages "in the vicinity of the Elkhorn" (2:623). In volume one, the authors write that the Omahas built a village, known in English as "The Little Village," on the Elkhorn River near Clark Creek in present Dodge County in the spring of 1841 when they were driven from the Missouri River by Sioux attacks. The Omahas lived there for two years before returning to "their old village on Omaha Creek, Dakota County, Nebraska" (1:100).
19. Gary E. Moulton, ed., *The Journals of the Lewis and Clark Expedition* (Lincoln: University of Nebraska Press, 1986), vol. 2, *August 30, 1803–August 24, 1804*, 419.
20. Ibid.
21. "Honorable A. W. Babbbitt and Company," *Frontier Guardian* (Kanesville [i.e., Council Bluffs], IA), 16 May 1851; "From the Plains," *Frontier Guardian* (Kanesville [i.e., Council Bluffs], IA), 13 June 1851; Sarah Hollister Harris, *An Unwritten Chapter of Salt Lake, 1851–1901* (New York: n.p., 1901), 5–31, available online at https://history.lds.org/overlandtravels/trailExcerptMulti?lang=eng&companyId=58&-sourceId=9288 (accessed 20 May 2015).
22. Lillian L. Fitzpatrick, *Nebraska Place-Names*, G. Thomas Fairclough, ed. (Lincoln: University of Nebraska Press, 1960), 56, 74, 78; C. H. Scoville, comp., *History of the Elkhorn Valley, Nebraska: An Album of History and Biography Containing a Descriptive Political, Statistical, Financial, Commercial, Agricultural, Educational, Religious and Biographical History and Reminiscences* (Chicago and Omaha: National Publishing Company, 1892; reproduced by Unigraphic, Inc., Evansville, IN, 1977), 184, 273, 278, 282.
23. J. W. C. Jones, "The Elkhorn Valley; its climate, soil, surface, water, timber grasses, products, people, and industrial resources; or, A month's talk, walk, and drive, in

the counties of Cuming, Stanton, Madison, Antelope, Wayne and Pierce, Nebraska" (West Point, NE: Republican Job Office, 1880), 1–3. Pages in the original are not numbered. Page numbers provided here are based on the assumption that the first interior page is page 1.

24. Ibid., 7–8.
25. Ibid., 10–11. Quotation on p. 10.
26. Ibid., 13–14.
27. Ibid., 14.
28. Ibid., 29.
29. Ibid., 10, 24.
30. Ibid., 20–22.
31. Frank Schmitt, "A Pioneer Nebraska Community," *Nebraska History* 18 (April–June, 1937): 80.
32. "The Late Chief of the Omaras," *Bangor (ME) Daily Whig and Courier*, 15 August 1855; Schmitt, 79–81.
33. Ella J. DeBell, "The Bruner and Neligh Families," *Nebraska History* 16 (April–June, 1935): 87.
34. Milton E. Holtz, "Early Settlement and Public Land Disposal in the Elkhorn River Valley, Cuming County, Nebraska Territory," *Nebraska History* 52 (Summer 1971), 115.
35. Thomas R. Buecker, *Water Powered Flour Mills in Nebraska* (Lincoln: Nebraska State Historical Society, 1983), 3–4; Grier, "The Elkhorn," 68.
36. Buecker, *Water Powered Flour Mills*, 24–26.
37. Ibid., 16–18.
38. DeBell, "The Bruner and Neligh Families," 89.
39. Jones, "The Elkhorn Valley," 36.
40. Scoville, *History of the Elkhorn Valley*, 191.
41. Sanborn Map Company, *Insurance Map of West Point, Nebraska* (New York: Sanborn Map Company, 1886, 1889, 1897, 1904, 1909, 1921); Buecker, *Water Powered Flour Mills*, 16–18, 28–30.
42. A. J. Leach, *A History of Antelope County, Nebraska, From Its First Settlement in 1868 to the Close of the Year 1883* (Chicago: The Lakeside Press, 1909), 12.
43. Ibid.
44. Scoville, *History of the Elkhorn Valley*, 188.
45. Ibid., 270.
46. Leach, *A History of Antelope County*, 94, 126–127.
47. Buecker, *Water Powered Flour Mills*, 30–31.
48. "Mill Dam over the Logan, Oakland, Neb.," photographic postcard, n.d., author's collection. The card bears a 1913 postmark, which indicates the image was made in or before 1913.

49. Leach, *A History of Antelope County*, 132–134.
50. *West Point (NE) Republican*, 24 April 1873.
51. Ibid., 29 May 1873.
52. Ibid., 5 June 1873.
53. Ibid., 12 June 1873.
54. The story of the aquarium car accident found wide distribution via the telegraph and versions appeared in newspapers in Bangor, Maine, San Francisco, and Macon, Georgia, to name a few examples. This telling relies on the following sources: *West Point (NE) Republican*, 12 June 1873; "Details of the Union Pacific Disaster at Elkhorn Bridge," *Daily Rocky Mountain News* (Denver, CO), 13 June 1873 [reprinted from the *Omaha Herald*]; "The Aquarium Car Wrecked," *Boston Daily Advertiser*, 14 June 1873; and, "The Lost Fish," *Daily Evening Bulletin* (San Francisco), 18 June 1873.
55. *West Point (NE) Republican*, 19 June 1873.
56. "Telegraphic," *Daily Rocky Mountain News* (Denver, CO), 31 March 1875; "High Water," *Daily Arkansas Gazette* (Little Rock), 14 April 1875; "Nebraska Notes," *St. Louis (MO) Globe-Democrat*, 10 May 1877; "Notes from Exchanges," *Daily Rocky Mountain News* (Denver, CO); 25 May 1877; Photograph of Norfolk, Nebraska, flood, catalog number 1986–3024–16, VF-1877 Flood, Elkhorn Valley Museum, Norfolk, NE; "Across the Continent: The Frank Leslie Excursion to the Pacific, from Omaha to Fremont and the Valley of the Platte," *Frank Leslie's Illustrated Newspaper* (New York), 1 September 1877.
57. *Fremont (NE) Weekly Herald*, 31 March 1881.
58. "The Flood," *West Point (NE) Republican*, 31 March 1881.
59. *Fremont (NE) Weekly Herald*, 21 April 1881; "Ousting Omaha," *Rocky Mountain News* (Denver, CO), 22 April 1881; "All About the Flood," *Rocky Mountain News* (Denver, CO), 23 April 1881; "The Missouri River," *Rocky Mountain News* (Denver, CO), 26 April 1881.
60. "The Flood," *West Point (NE) Republican*, 31 March 1881.
61. *Fremont (NE) Weekly Herald*, 31 March 1881.
62. "The Freshet," *Neligh (NE) Leader*, 1 June 1888; "The Freshet," *Neligh (NE) Leader*, 8 June 1888. Collections of lower-basin newspapers are incomplete for May 1888; thus, the discussion is limited to the upper basin.
63. "The Freshet," *Neligh (NE) Leader*, 8 June 1888.
64. "Raging Waters," *Norfolk (NE) Daily News*, 29 May 1888; "The Booming North Fork," *Norfolk (NE) Daily News*, 30 May 1888; *Norfolk (NE) Daily News*, 31 May 1888; *Norfolk (NE) Journal*, 31 May 1888; "The Flood," *Norfolk (NE) Daily News*, 31 May 1888.
65. *Neligh (NE) Leader*, 1 June 1888.
66. "The Flood," *Norfolk (NE) Daily News*, 31 May 1888; "The Flood," *Norfolk (NE) Daily News*, 7 June 1888.

67. Scoville, *History of the Elkhorn Valley, Nebraska*, 179. Scoville reports flooding in July 1891 that exceeded the June floods. The July editions of the *Cuming County (NE) Advertiser* do not corroborate Scoville's assertion that the depth and damages of the July floods exceeded those of the June flood. Those editions report rainstorms during the period Scoville mentions, but no flooding of the river. The *Advertiser* reported that, according to the oldest settlers, the June rains and floods were the greatest since those of 1881. "The Flood," *Cuming County (NE) Advertiser*, 30 June 1891.
68. Buecker, *Water Powered Flour Mills in Nebraska*, 45–48.
69. Scoville, *History of the Elkhorn Valley, Nebraska*, 246, 276.

Chapter 3

1. "A View of the Elkhorn, Neligh, Neb.," postcard, n.d. [the card bears a 1907 postmark], author's collection. For a discussion of work as a way to know a river, see Richard White, *The Organic Machine* (New York: Hill and Wang, 1995).
2. "Neligh Mills and Dam, Neligh, Neb.," hand-tinted postcard, n.d. [image appears on the cover of *Nebraska History*, 79 (Winter 1998), which dates the image to about 1907; postcard in author's possession bears a 1911 postmark], author's collection.
3. Kent E. Martin, *Neligh Mill State Historic Site, Neligh, Nebraska, Nebraska State Historical Society: Self-Guided Tour* (Lincoln: Nebraska State Historical Society, 1997), 5–6, 13, 16–17; Buecker, *Water Powered Flour Mills in Nebraska*, 46, 65–66; "'So-Lite': The Perfect Flour," advertisement for Neligh Mill, *Norfolk (NE) Daily News*, 8 May 1914.
4. Martin, *Neligh Mill State Historic Site*, 18–19; Buecker, *Water Powered Flour Mills in Nebraska*, 65–66; "Edison Gem Lamp," advertisement, Neligh Mill Light Department, n.d., on display at Neligh Mill State Historic Site; "A Good Electric Flat Iron," advertisement for Triangle Lektrik Iron, Neligh Mill Light Department, 1913, on display at Neligh Mill State Historic Site; "Press the Button—Turn Work into Play," advertisement for Frantz Premier Electric Cleaner, 1915, on display at Neligh Mill State Historic Site.
5. "Norfolk Is Flooded," *Norfolk (NE) Daily News*, 20 March 1897; "Water Is Going Down," *Norfolk (NE) Daily News*, 22 March 1897.
6. "Vastly Too Much Wet," *Norfolk (NE) Daily News*, 6 April 1897.
7. Klein and Zellmer, *Mississippi River Tragedies*, 39–41, 42–43.
8. Ibid., 28–29; quotation on p. 28.
9. U.S. Army Corps of Engineers, Omaha District, *Review Report on Improvement for Flood Control and Allied Purposes, Elkhorn River and Its Tributaries, Nebraska* (Omaha: Department of War, 1947), 44.
10. Ibid., 44.
11. Ibid., 45–46.
12. Klein and Zellmer, *Mississippi River Tragedies*, 44; "Rivers are Rising Fast," and "The Winter of Big Snows," *Norfolk (NE) Daily News*, 25 March 1912.

13. Bomer, *Sons of the River*, 76; Emanuel Wolfe, ["Cutting blocks of ice from the Elkhorn River?], glass plate negative, n.d., RG2836.PH0-000293, Nebraska State Historical Society; Esther Kolterman Hansen, *Along Pioneer Trails in Pierce County, Nebraska* (n.p., n.d., 1940), 77 [available at local history library, Nebraska State Historical Society]; "Northfork River Highest in Years," *Norfolk (NE) Daily News*, 28 March 1912.
14. U.S. Engineer Office, *Interim Report on Flood Control for the City of Norfolk, Elkhorn River Basin, Nebraska* (Omaha: U.S. Engineer Office, 1946), 26.
15. "Again Out of Water," *Norfolk (NE) Daily News*, 1 April 1912.
16. "The Elkhorn Is Falling," *Norfolk (NE) Daily News*, 2 April 1912; Norfolk Flood, 1912, photograph, catalog number 7777-51-08 VF-1912 Flood, Elkhorn Valley Museum, Norfolk, NE.
17. "Again Out of Water," 1 April 1912; "Madison City Was Hit Hard," *Norfolk (NE) Daily News*, 2 April 1912; "Place Loss at $25,000," *Norfolk (NE) Daily News*, 2 April 1912.
18. "Elkhorn Bridge Damaged," *West Point (NE) Republican and Cuming County Advertiser*, 12 April 1912.
19. Ibid.
20. "Expected Flood Arrives," *West Point (NE) Republican and Cuming County Advertiser*, 5 April 1912.
21. "Water Pumping Solved," *West Point (NE) Republican and Cuming County Advertiser*, 17 May 1912.
22. "Expected Flood Arrives," 5 April 1912.
23. Brian P. Dunnigan, Head, Flood Plain Management, State of Nebraska Natural Resources Department, interview with the author, Lincoln, 11 February 1992.
24. "High Water," *West Point (NE) Republican and Cuming County Advertiser*, 12 April 1912.
25. Klein and Zellmer, *Mississippi River Tragedies*, 55–56.
26. Quoted in ibid., 57.
27. Ibid., 56–58; quotation on p. 57.
28. Buecker, *Water Powered Flour Mills*, 48; U.S. Army Corps of Engineers, Omaha District, 46; Bentall, *Water Supplies and the Land*, 19.
29. "Big Flood Is On Way Here from Pierce," *Norfolk (NE) Daily News*, 22 March 1917; "Floods Follow Torrential Rain," *Norfolk (NE) Daily News*, 28 May 1917; "Water Falls After Heavy Flood Loss," *Norfolk (NE) Daily News*, 26 March 1917; "Raising Dykes to hold back flood, Norfolk Nebr. Mar. 23, 1917," photograph, catalog number 2001-84-76 VF-1917, Elkhorn Valley Museum, Norfolk, NE.
30. "Floods Follow Torrential Rain," *Norfolk (NE) Daily News*, 28 May 1917; Buecker, *Water Powered Flour Mills*, 65–66.
31. "Near Death at Battle Creek in Flooded River," *Norfolk (NE) Daily News*, 29 May 1917.

32. "Elkhorn Flood Covers Fields, Damages Crops," *Norfolk (NE) Daily News*, 30 May 1917.
33. "Flood Damages Pilger Stores," *Norfolk (NE) Daily News*, 5 June 1917; "Lives Lost in Storm at Gretna; Tracks Flooded," *Norfolk (NE) Daily News*, 6 June 1917.
34. Klein and Zellmer, *Mississippi River Tragedies*, 59, 66–67, 76.
35. "Waters Cover the Earth," *West Point (NE) Republican*, 30 April 1920.
36. "Damage by Flood Feared as River Overflows Bank," *Norfolk (NE) Daily News*, 20 April 1920; Martin, *Neligh Mill Museum*, 26; Buecker, *Water Powered Flour Mills*, 65–66.
37. "Waters Cover the Earth."
38. "Flood Undermines Three Spans of Bridge, Cuts Off Train Service with West," *Norfolk (NE) Daily News*, 26 April 1920.
39. "Stores at Beemer Invaded by Flood," *Norfolk (NE) Daily News*, 26 April 1920.
40. "Tilden Man Drowned in Flooded Elkhorn," *Norfolk (NE) Daily News*, 29 April 1920.
41. "30 Marooned as River Flood Forms Island," *Norfolk (NE) Daily News*, 30 April 1920.
42. "Highway Through Holt Being Made Safe From Flood," *Norfolk (NE) Daily News*, 7 May 1920.
43. "The Proposed Elkhorn Drainage District," *West Point (NE) Republican*, 25 June 1920.
44. Floodplain and Dam Safety Division, *Flood Hazard Mitigation Plan, State of Nebraska* (Lincoln: Nebraska Department of Natural Resources, 2013), Appendix E, "Nebraska Flooding History Summary."
45. Stacey Stubbs, "'Gentle River Goes Mad': The Republican River Flood of 1935 and Its New Deal Legacy," *Nebraska History* 97 (Spring 2016): 2–15.
46. Klein and Zellmer, *Mississippi River Tragedies*, 59, 66–67, 76.
47. U.S. Army Corps of Engineers, Omaha District, 45–46; Klein and Zellmer, *Mississippi River Tragedies*, 81.
48. U.S. Army Corps of Engineers, Omaha District, 45–46.

Chapter 4

1. "Water Rising at Pierce; Madison, Battle Creek Hit," "Spillway Helps Prevent Flood in this City," *Norfolk (NE) Daily News*, 4 June 1940.
2. "Water Rising at Pierce; Madison, Battle Creek Hit"; "Worst Flood in History Inundates Much of Cuming County," *West Point (NE) Republican*, 6 June 1940.
3. "Two Dead, 6 Missing in Flood," *Fremont (NE) Daily Tribune*, 4 June 1940.
4. U.S. Army Corps of Engineers, Omaha District, 45–46.
5. "Railroad Service on Regular Schedule," "Emergency Work Opens Bridges," *West Point (NE) Republican*, 20 June 1940.

6. “Water Rising at Pierce; Madison, Battle Creek Hit”; “Many Towns Are Lacking Rail Service.”
7. “Farmland in County Flooded,” *Fremont (NE) Daily Tribune*, 5 June 1940.
8. “Water Rising at Pierce; Madison, Battle Creek Hit”; “Many Towns are Lacking Rail Service.”
9. “Water Rising at Pierce; Madison, Battle Creek Hit.”
10. “Farmers Find Flood Damage Is Tremendous,” *Norfolk (NE) Daily News*, 5 June 1940.
11. “Pilger Flooded,” *Norfolk (NE) Daily News*, 6 June 1940.
12. “Red Cross Closes Emergency Relief Station,” *West Point (NE) Republican*, 13 June 1940; “150 Homeless in West Point Region,” *Fremont (NE) Daily Tribune*, 5 June 1940. These accounts disagree on the number of people receiving aid from the Red Cross. The former reports food and shelter given to twenty families or seventy-five people. The latter reports assistance given to 150 people.
13. “Red Cross Opens Local Relief Headquarters,” *West Point (NE) Republican*, 6 June 1940; “150 Homeless in West Point Region.”
14. “Worst Flood in History Inundates Much of Cuming County.”
15. “3 Inches More Rain Adding to Flood Worries,” *Norfolk (NE) Daily News*, 6 June 1940; “Worst Flood in History Inundates Much of Cuming County.”
16. “3 Inches More Rain Adding to Flood Worries”; “New Floods Threaten Pilger, Wisner,” *Fremont (NE) Daily Tribune*, 6 June 1940.
17. “Pilger Flooded,” *Norfolk (NE) Daily News*, 6 June 1940.
18. “Waters Are High in Scribner Vicinity,” *Norfolk (NE) Daily News*, 8 June 1940.
19. “Experts Advise Crops for Area of Flood,” *Fremont (NE) Daily Tribune*, 6 June 1940.
20. “AAA Regulations Are Modified,” *Fremont (NE) Daily Tribune*, 7 June 1940.
21. “Red Cross Opens Local Relief Headquarters,” *West Point (NE) Republican*, 6 June 1940.
22. “Board Seeks Bridge Repair Funds,” *West Point (NE) Republican*, 13 June 1940.
23. Cuming County, Nebraska, Board of Supervisors, *County Board Proceedings*, 7 June 1940, 7:457–458.
24. “Emergency Work Opens Bridges,” *West Point (NE) Republican*, 20 June 1940.
25. “Railroad Service on Regular Schedule,” *West Point (NE) Republican*, 20 June 1940.
26. Cuming County, Nebraska, Board of Supervisors, 2 July 1940, 7:459.
27. Cuming County, Nebraska, Board of Supervisors, 12 August 1940, 7:475. Taxing entities sometimes set levies using a mill rate calculation. In the simplest terms, one mill equals 1/1000 of a dollar or one dollar in tax for every one thousand dollars of assessment. Using six as the mill rate, a taxpayer with property valued at $50,000 would owe $300 in taxes.
28. Cuming County, Nebraska, Board of Supervisors, 19 August 1940, 8: 53.
29. “Ag Official Points to Flood Control Work,” *West Point (NE) Republican*, 11 July 1940.

30. U.S., House of Representatives, 73d Cong., 2d sess., 1934, Doc. 197, *Platte River, Colo., Wyo., and Nebr.*, serial 9826, 359–362.

31. U.S., House of Representatives, 81st Cong., 1st sess., 1949, Doc. 215, *Elkhorn River and Tributaries, Nebraska*, Serial 11324, 15; U.S. Army Corps of Engineers, Omaha District, *Supplement to Detailed Project Report: Flood Protection Project, Pebble Creek and Elkhorn River at Scribner, Nebraska* (Omaha: U.S. Army Corps of Engineers, 1991), 5; U.S. Engineer Office, *Interim Report on Flood Control for the City of Norfolk, Elkhorn River Basin, Nebraska* (Omaha: U.S. Engineer Office, 1946), 15; "Creek Near City Helped Cause Flood," *Norfolk (NE) Daily News*, 16 May 1944.

32. "Elkhorn Valley Hit by Floods, Norfolk Stores Suffer Damage," *Neligh (NE) Leader*, 17 May 1944.

33. U.S. Army Corps of Engineers, Omaha District, *Review Report on Improvement for Flood Control and Allied Purposes, Elkhorn River and Its Tributaries, Nebraska, Appendix VIII, Transcript of Public Hearings* (Omaha: War Department, 1947), 154.

34. "Flood Leaves 300 to 400 Homeless," *Norfolk (NE) Daily News*, 13 May 1944.

35. Ibid.

36. *Interim Report on Flood Control for the City of Norfolk*, 15.

37. "Flood Leaves 300 to 400 Homeless."

38. "City Is Making Rapid Strides Toward Recovery," *Norfolk (NE) Daily News*, 15 May 1944; "Norfolk Nebr Flood 5-12-44," photograph, Elkhorn Valley Museum, Norfolk, NE.

39. "City Is Making Rapid Strides Toward Recovery"; "Elkhorn Valley Hit by Floods, Norfolk Suffers Damage."

40. *Interim Report on Flood Control for the City of Norfolk*, 15.

41. "Warnings Are Issued About Water, Food," *Norfolk (NE) Daily News*, 15 May 1944; "City Is Making Rapid Strides Toward Recovery"; *Kreiter Directory Co's. Norfolk City and Madison County Directory, 1944–45*, vol. XV (Rochester, MN, and Norfolk, NE: Kreiter Directory Co., 1944), 72, 515, 544.

42. "Warnings Are Issued about Water, Food"; "Workers Are Badly Needed by Merchants," *Norfolk (NE) Daily News*, 16 May 1944; "Health Board Orders Stores to Thoroughly Clean Out Basements," *Norfolk (NE) Daily News*, 19 May 1944; "50 Volunteers from Other Towns Help in Flood Debris Clean-up," *Norfolk (NE) Daily News*, 22 May 1944.

43. U.S. Engineer Office, *Interim Report on Flood Control for the City of Norfolk, Elkhorn River Basin, Nebraska* (Omaha: U.S. Engineer Office, 1946), 17.

44. Ibid., 17–18.

45. "Flood Waters Damage Much Farm Property," *Stanton (NE) Register*, 18 May 1944.

46. "Flood Waters Do Not Reach 1940 Level at West Point," *Norfolk (NE) Daily News*, 15 May 1944; "Flood Waters May Next Hit at Arlington," *Norfolk (NE) Daily News*, 15 May 1944; "Flood Damages West Part of Town," *West Point (NE) Republican*, 18 May 1944.

47. "Flood Damages West Part of Town."
48. "Elkhorn River, Logan Creek Rise," *Hooper (NE) Sentinel*, 18 May 1944.
49. "Elkhorn Floods 5,000 Acres Survey Shows," *Fremont (NE) Guide and Tribune*, 17 May 1944.
50. House of Representatives, *Elkhorn River and Tributaries*, 15; U.S. Army Corps of Engineers, Omaha District, *Flood Protection Project: Pebble Creek and Elkhorn River at Scibner, Nebraska; Supplement to Detailed Project Report* (Omaha: U.S. Army Corps of Engineers, 1991), 5.
51. House of Representatives, *Elkhorn River and Tributaries*, 15.
52. Ibid., 20–21; "Elkhorn, Tributaries Hit Area with 'Worst Flood' in History," *Fremont (NE) Guide and Tribune*, 12 June 1944; "Scribner, Howells and Hooper Badly Flooded," *Norfolk (NE) Daily News*, 12 June 1944.
53. House of Representatives, *Elkhorn River and Tributaries*, 15; "Scribner, Howells, and Hooper Badly Flooded," "Damage Placed at Millions of Dollars," *Norfolk (NE) Daily News*, 12 June 1944.
54. "Flash Flood Sunday Drowns Livestock, Ruins Crops, Roads," *Stanton (NE) Register*, 15 June 1944; "Losses Reported from Area Floods," *Stanton (NE) Register*, 15 June 1944.
55. "11–14–Inch Rains South of Pilger Damage Crops," *Stanton (NE) Register*, 15 June 1944.
56. "Flood Waters Do Extensive Damage," *West Point (NE) Republican*, 15 June 1944; "Damage by Elkhorn Defies Estimation," *Fremont (NE) Guide and Tribune*, 13 June 1944. The former says the flood warning sounded at 7 a.m., while the latter reports the whistle sounding at 7:30 a.m.
57. "Damage by Elkhorn Defies Estimation," "Scribner, Howells, and Hooper Badly Flooded," "Flood Waters Do Extensive Damage," *West Point (NE) Republican*, 15 June 1944.
58. "Scribner, Howells, and Hooper Badly Flooded."
59. "Maple Creeks Take Heavy Property Toll," *Fremont (NE) Guide and Tribune*, 13 June 1944.
60. "Creek Becomes Raging Torrent, Rips Destructive Path through Clarkson," *Norfolk (NE) Daily News*, 12 June 1944; "Wanek Family Says Flood Like 'White Cloud,'" *Fremont (NE) Guide and Tribune*, 13 June 1944. Reports disagree about the flood's timing. The consensus seems to be that the "wall of water" descended around 2 a.m.
61. U.S. Army Corps of Engineers, Omaha District, *Detailed Project Report: Flood Protection Project at Scribner, Pebble Creek at Scribner, Nebraska* (Omaha: U.S. Army Corps of Engineers, 1985), 11, 13, 17; U.S. Army Corps of Engineers, Omaha District, *Supplement to Detailed Project Report*, 5–6.
62. "Hooper and Area Hit By Damaging Flood Waters," *Hooper (NE) Sentinel*, 15 June 1944.

63. "Elkhorn, Tributaries Hit Area with 'Worst Flood' in History," *Fremont (NE) Guide and Tribune*, 12 June 1944.
64. "Maple Creeks Take Heavy Property Toll."
65. "Scribner... Flooded," "Elkhorn... 'Worst Flood' in History," "Waterloo People Return to Homes as Waters Fall," *Fremont (NE) Guide and Tribune*, 16 June 1944.
66. "Flood Inundates 92,480 Acres, Damages 750 Homes in County," *Fremont (NE) Guide and Tribune*, 17 June 1944; "Seek Outside Aid for Flood Cleanup," *Fremont (NE) Guide and Tribune*, 24 June 1944.
67. "Red Cross Will Give Financial Help to Needy," *Fremont (NE) Guide and Tribune*, 15 June 1944; "Hooper Could Use Volunteer Workers to Remove Debris," *Fremont (NE) Guide and Tribune*, 18 June 1944.
68. "Seed Corn Beetles Form New Threat to County Crop," *Fremont (NE) Guide and Tribune*, 24 June 1944.
69. "Elkhorn... 'Worst Flood' in History."
70. "State Asks $212,500 for Repair of Flooded Roads," *Norfolk (NE) Daily News*, 10 July 1944.
71. "Scribner... Flooded."
72. "Rebuilding Oakdale Line to Begin Soon," *Norfolk (NE) Daily News*, 9 September 1944.
73. "Wisner Farmer, Flooded Out, Kills Himself," *Fremont (NE) Guide and Tribune*, 14 June 1944.
74. "Guard Mustered Out Following Week of Duty in Flood Area," *Fremont (NE) Guide and Tribune*, 17 June 1944.
75. "Dike Repair Work on Northfork Is Ordered," *Norfolk (NE) Daily News*, 20 May 1944.
76. "Scribner... Flooded," *Norfolk (NE) Daily News*, 12 June 1944.
77. "Red Cross Opens Rehabilitation Headquarters," *Norfolk (NE) Daily News*, 15 May 1944; "800 Typhoid Shots Have Been Given," *Norfolk (NE) Daily News*, 15 May 1944; "Red Cross Helps More People Here in Past Week Than Ever in History," *Norfolk (NE) Daily News*, 22 May 1944.
78. "Red Cross Will Give Financial Help to Needy," *Fremont (NE) Guide and Tribune*, 15 June 1944.
79. "Scribner... Flooded"; "Elkhorn... 'Worst Flood' in History"; "Maple Creeks Take Heavy Property Toll"; "Residents Estimate Scribner Cleanup Will Take Weeks," *Fremont (NE) Guide and Tribune*, 18 June 1944.
80. "Hooper... Hit by... Flood Waters," *Hooper (NE) Sentinel*, 15 June 1944; "Special Flood Notice," *Hooper (NE) Sentinel*, 22 June 1944.
81. "Scribner... Flooded."
82. "Damage... Defies Estimation."
83. "No Aid Received at Snyder," *Fremont (NE) Guide and Tribune*, 13 June 1944.

84. Cuming County, Nebraska, Board of Supervisors, *County Board Proceedings*, 10 August 1944, 8:276–277.
85. "Gross Says Corn Up, Not Covered with Mud, May Live If Drained Quickly," *Fremont (NE) Guide and Sentinel*, 14 June 1944. Agriculturally minded readers might wish to know that Gross recommended Iowa 939 or Iowa 306 corn for replanting.
86. U.S. Army Corps of Engineers, Omaha District, *Review Report on Improvement for Flood Control and Allied Purposes, Elkhorn River and Its Tributaries, Nebraska* (Omaha: War Department, 1947), 98.
87. *Platte River*, 359–362; Marian E. Ridgeway, *The Missouri Basin's Pick-Sloan Plan: A Case Study in Congressional Policy Determination*, Illinois Studies in the Social Sciences, vol. 35 (Urbana: University of Illinois Press, 1955), 35.
88. U.S., House of Representatives, *Flood Control Act of 1936, Statutes at Large*, vol. 49, part 1, secs. 1–3, 1570–1571 (1936).
89. Ridgeway, 3–16; Michael L. Lawson, *Dammed Indians: The Pick-Sloan Plan and the Missouri River Sioux, 1944–1980* (Norman: University of Oklahoma Press, 1982), 11–15; Klein and Zellmer, *Mississippi River Tragedies*, 66–67, 76, 81.
90. *Elkhorn River and Tributaries*, 2.
91. *Review Report... Transcript of Public Hearings*, 22; "Freeman Is Told of Flood Damage," *Fremont (NE) Guide and Tribune*, 30 August 1944.
92. 'Meeting Is Held Flood Problems," *Hooper (NE) Sentinel*, 29 June 1944.
93. *Review Report... Transcript of Public Hearings*, 28.
94. Ibid., 35–36.
95. Ibid., 280; "Beemer Estimates $400,000 Flood Loss," *West Point (NE) Republican*, 31 August 1944.
96. *Review Report... Transcript of Public Hearings*, 42–43, 61–65.
97. "Freeman... Told of... Damage."
98. "Sponging Up the Elkhorn," *Fremont (NE) Guide and Tribune*, 1 September 1944.
99. House of Representatives, *Elkhorn River and Tributaries*, 34.
100. "No Funds Available for Elkhorn Work," *West Point (NE) Republican*, 13 July 1944.
101. House of Representatives, *Elkhorn River and Tributaries*, 54–55.
102. *Review Report*, 58.
103. Ibid., 68.
104. Ibid., 70–71.
105. Ibid., 72–73.
106. House of Representatives, *Elkhorn River and Tributaries*, 48–54.
107. *Review Report*, 1946, 59.
108. *Interim Report on Flood Control for the City of Norfolk*, 21; *Review Report*, 59–62.
109. *Review Report*, 59–62.
110. Ibid., 62.
111. Ibid., 63–64, 66.

112. Ibid., 64–67; House of Representatives, *Elkhorn River and Tributaries*, 50–54.
113. House of Representatives, *Elkhorn River and Tributaries*, 74, 98, 101, 104.
114. Ibid., 35.
115. *Review Report*, 52–53.
116. Dunnigan, interview.
117. *Review Report*, 67, 69, 71, 73.
118. U.S. Army Corps of Engineers, Omaha District, *Review Report on Improvement for Flood Control and Allied Purposes, Elkhorn River and Its Tributaries, Nebraska* (Omaha: War Department, 1947), 101–102.
119. House of Representatives, *Elkhorn River and Tributaries*, 59–60, 67.
120. Ibid., 59–60; *Review Report... Transcript of Public Hearings*, 136.
121. *Review Report... Transcript of Public Hearings*, 149.
122. Donald Worster, *Dust Bowl: The Southern Plains in the 1930s* (New York: Oxford University Press, 1979), 4; Worster, *Rivers of Empire*, 332.
123. *Review Report*, 101.
124. Nebraska, Natural Resources Commission, *Flood Plain Study: Elkhorn River* (Lincoln: Nebraska Natural Resources Commission, 1987), vol. 1, *Cuming County, Nebraska*, 1. The 1983 Nebraska Flood Plain Management Act directed the Nebraska Natural Resources Commission to make maps to develop or support flood plain regulation by local governments.

Chapter 5

1. "'Mean Elkhorn River' Gobbles Up Valley Family's Land," *Omaha (NE) World-Herald*, 18 June 1984.
2. U.S. Army Corps of Engineers, Omaha District, *Project Maps, Omaha District: Status of River and Harbor and Flood Control Projects* (Omaha: Corps of Engineers, 1983), 71–72, 74, 76, 83, 87.
3. Ibid., 73–84.
4. Ibid., 75, 77, 80, 86, 89.
5. "Flood Fears Rise Along Eastern Rivers," *Lincoln (NE) Journal Star*, 7 March 2010.
6. "Trees Will Be Removed From River," *Norfolk (NE) Daily News*, 14 April 1947.
7. "Flash Flood Strikes Dodge Area; C.&N.W. Tracks Are Damaged," *Norfolk (NE) Daily News*, 15 May 1947.
8. "Flash Floods Strike 2 Towns; C. and N.W. Train Leaps Tracks," *Norfolk (NE) Daily News*, 2 June 1947.
9. "C&NW Passenger Train Is Derailed Near Nickerson; Elkhorn Valley Is Soaked," *Fremont (NE) Guide and Tribune*, 2 June 1947; "Flood Warnings Are Given Residents in Elkhorn River Area," *Fremont (NE) Guide and Tribune*, 3 June 1947.
10. "Drainage Ditch Work to Start Sometime in June," *Stanton (NE) Register*, 5 June 1947.

11. "Flash Flood Strikes Dodge; High Waters Hit Other Sections," *Norfolk (NE) Daily News*, 12 June 1947; "Two-State Wind Storm Kills Three Causes Heavy Damage," *Cuming County (NE) Democrat*, 12 June 1947.
12. "New Elkhorn Path; Iowa Threats Eased," *Fremont (NE) Guide and Tribune*, 16 June 1947; "High Water Takes Bridge," *Norfolk (NE) Daily News*, 16 June 1947; "Roaring Elkhorn Takes Out Ertzner Bridge," *Stanton (NE) Register*, 19 June 1947;
13. "Elkhorn River Floods Lowlands Near Neligh," *Neligh (NE) Leader*, 18 June 1947.
14. "Union Creek Area Has Great Damage," *Stanton (NE) Register*, 19 June 1947.
15. "Floods Hit Inman, Page, Ewing Areas; Bridges Ripped Out," *Norfolk (NE) Daily News*, 23 June 1947; "Elkhorn Overflows Near Meadow Grove; River Rising Here," *Norfolk (NE) Daily News*, 24 June 1947.
16. "Elkhorn River Goes on Rampage Flooding Valley; Cuts Highway 14," *Neligh (NE) Leader*, 25 June 1947; "Highway 14 South From Neligh Open to Traffic," *Neligh (NE) Leader*, 9 July 1947.
17. "Several Families at Beemer Evacuated as Elkhorn Still Rises," *Norfolk (NE) Daily News*, 26 June 1947; "Flood Crest Here within 48 Hours," *Cuming County (NE) Democrat*, 26 June 1947; "Beemer, West Point Flood Stage Is Near," *Fremont (NE) Guide and Tribune*, 26 June 1947; "Flood Damages Elkhorn Bridge on No. 81 Here," *Norfolk (NE) Daily News*, 27 June 1947; "Flood Waters Pound C. & N.W. Tracks, Beemer," *Norfolk (NE) Daily News*, 28 June 1947; "Flood Waters Recede; West Point, Scribner," *Norfolk (NE) Daily News*, 30 June 1947.
18. "Co. Commissioners Estimate Flood Damage at $85,000," *Stanton (NE) Register*, 3 July 1947.
19. "Minor Flood Knocks Out Road Bridge," *Norfolk (NE) Daily News*, 10 June 1960 ("minor rampage").
20. "NE Nebraska Towns in Path of Flood Waters," *Norfolk (NE) Daily News*, 20 June 1960; "Northwest Part of Madison Floods," *Norfolk (NE) Daily News*, 20 June 1960; "Storms Effects Slowing Traffic," *Norfolk (NE) Daily News*, 20 June 1960; "Latest Elkhorn Flood Woes Appear Ending," *Norfolk (NE) Daily News*, 21 June 1960; "Battle Creek Starts Messy Job of Cleanup," *Norfolk (NE) Daily News*, 21 June 1960; "'Copters to Be Used in Search," *Norfolk (NE) Daily News*, 21 June 1960; "County Is Declared Disaster Area," *Norfolk (NE) Daily News*, 22 June 1960; "Search for Body Continues to Be Fruitless," *Norfolk (NE) Daily News*, 22 June 1960.
21. "More Lessons on Flood Control," *Norfolk (NE) Daily News*, 23 June 1960.
22. "Northeast Nebraska Alert as Waters Rise," *Norfolk (NE) Daily News*, 24 March 1962; "Northfork, Elkhorn to Run Near Flood Stage," *Norfolk (NE) Daily News*, 26 March 1962.
23. "Water Continues to Rise Along Northfork," *Norfolk (NE) Daily News*, 27 March 1962; "Flood Crests Here Tonight," *Norfolk (NE) Daily News*, 28 March 1962; "Area Flood Meeting Wednesday," *Norfolk (NE) Daily News*, 31 March 1962.

24. "Northfork Begins Slow Drop from Night High," *Norfolk (NE) Daily News*, 29 March 1962; "It Happened Before—50 Years Ago," *Norfolk (NE) Daily News*, 29 March 1962; "Pierce, West Point May Attain Dry 1963," *Norfolk (NE) Daily News*, 28 March 1962; "Area Flood Meeting Wednesday," *Norfolk (NE) Daily News*, 31 March 1962.
25. Hooper Mrs. Jaycees, *Centennial History of Hooper, 1876—1976* (N.p.: Hooper Centennial-Bicentennial Committee, 1976), 123–125.
26. Ibid., 146–147. According to the book, the Corps turned over the dike to the village on November 15, 1965, which seems to be a typographical error.
27. "Rain and Snow Melt Cause Area Flooding," *Norfolk (NE) Daily News*, 9 February 1966.
28. Ibid.; "Flood Moves Downstream; Morrison Asks for Conservation," *Norfolk (NE) Daily News*, 11 February 1966.
29. "Ice Jam Breaks, Families Return Home," *Norfolk (NE) Daily News*, 10 February 1966; "Omaha Time Capsule," *Omaha (NE) World-Herald*, 12 February 2013 [reprint of news story from 12 February 1966].
30. "Another Flood Reminder," *Norfolk (NE) Daily News*, 10 February 1966; "Flood Moves Downstream; Morrison Asks Conservation," *Norfolk (NE) Daily News*, 11 February 1966.
31. "Flash Flooding Spills Water in Low Areas," *Norfolk (NE) Daily News*, 25 June 1969.
32. "Waters Recede; Crane Tries to Break Log Jam," *Norfolk (NE) Daily News*, 26 June 1969.
33. Nebraska Department of Natural Resources, "Small Watersheds Flood Control Fund," (http://dnr.nebraska.gov/small-watersheds-flood-control-fund, accessed 12 August 2015). See also Nebraska Department of Natural Resources, *Small Watersheds Flood Control Fund Brochure* (June 2009), available as pdf via hyperlink at the website.
34. Nebraska Soil and Water Conservation Commission, *Flood Prevention and Damage Reduction*, State Water Plan, Publication Number 401, Revised Printing (Lincoln: The Commission, 1968), 1. Hereafter cited as NSAWCC 401.
35. Ibid., transmittal letter.
36. Ibid., 1.
37. Ibid., 2.
38. Ibid.
39. Ibid., 3.
40. Ibid., 6.
41. Ibid., 10–12.
42. Ibid., Foreword.
43. Klein and Zellmer, *Mississippi River Tragedies*, 112–113.
44. Douglas Helms, "Getting to the Roots," *People Protecting Their Land: Proceedings, Volume 1*, 7th ISCO Conference (Sydney, Australia: International Soil Conservation

Organization, 1992), 299–301, accessed 24 September 2015, http://www.nrcs.usda.gov/wps/portal/nrcs/detail/national/about/history/?&cid=nrcs143_021394; Nebraska Association of Resource Districts, "About NRDs," accessed 12 August 2015, http://nrdnet.org/about-nrds.php; Lower Platte North Natural Resource District, "About Us," accessed 12 August 2015, http://www.lpnnrd.org/about/history.html; Nebraska Association of Resource Districts, "Find Your NRD," accessed 12 August 2015, http://nrdnet.org/find-your-nrd.php. For a thorough legislative history on the creation of Nebraska's NRDs, see Hazel M. Jenkins, *A History of Nebraska's Natural Resource Districts* (Lincoln: Nebraska Department of Natural Resources, 1975; revised 2009), accessed 29 September 2015, http://dnr.ne.gov/nebraska-s-natural-resources-districts-history-updated-july-2009. See also Ann Bleed, Jim Barr, Dayle Williamson, and Gayle Starr, "The Creation of Nebraska's Natural Resource Districts," *Nebraska History* 97 (Winter 2016): 199–214.

45. The 1970s were characterized as "dry years" in Bleed, Barr, Williamson, and Starr. Ibid., 208. For information about Hooper, see Hooper Mrs. Jaycees, *Centennial History of Hooper*, 125.
46. "Northfork under Control," *Norfolk (NE) Daily News*, 21 March 1978; "7 Counties Affected; One Death," *Norfolk (NE) Daily News*, 20 March 1978; "Damaging Losses," *Norfolk (NE) Daily News*, 24 March 1978.
47. "Flood Damage Totals More Than $600,000 on State Spans, Roads," *Omaha World-Herald*, 19 June 1984.
48. "'Mean Elkhorn River' Gobbles Up Valley Family's Land," *Omaha (NE) World-Herald*, 18 June 1984; "Flood Victim Relief Funds Hit $55,953," *Omaha World-Herald*, 19 June 1984.
49. "Elkhorn River Menaces Home, Puts Couple in Quandary," *Omaha World-Herald*, 22 June 1984.
50. Ibid.
51. For a discussion of a river's flood plain as a natural safety valve and the problems of building in the Missouri River's flood plain, see Steinberg, *Acts of God*, 97–115.
52. Fred Thomas and Melinda Motty, "Rising Floodwaters Force More to Flee," *Omaha (NE) World-Herald*, 18 June 1990; Cindy Connolly, "Sandbaggers Keep Working Northeast Nebraska Cleaning Up after Flooding," *Omaha (NE) World-Herald*, 18 June 1990.
53. "Rising Floodwaters Force More to Flee"; "Sandbaggers Keep Working Northeast Nebraska Cleaning Up after Flooding"; Melinda Mooty and Fred Thomas, "As Flooding Wanes, Damages Grow," *Omaha (NE) World-Herald*, 20 June 1990; "West Point Death Is Blamed on Flood," *Omaha (NE) World-Herald*, 19 June 1990.
54. Julie Anderson, "Flood Ingredients Began to Jell a Month Earlier," *Omaha World-Herald*, 21 March 1993; Fred Thomas, "Flood Forces 1,400 to Evacuate; Traffic Rerouted as Overflow Closes I-10 and Highway 6," *Omaha (NE) World-Herald*, 11 March 1993.

55. Cindy Connolly, "Nebraska Efforts Focus on Worst Elkhorn, Loup and Platte," *Omaha (NE) World-Herald*, 10 March 1993; Greg Wees, "Area Rivers, Creeks Overflowing," *Norfolk (NE) Daily News*, 10 March 1993; Greg Wees, "Threat of Flood Wanes in Area," *Norfolk (NE) Daily News*, 11 March 1993.
56. Wees, "Threat of Flood Wanes in Area."
57. "Laura Erbst and Kent Warneke, "Area Towns Are Battling High Waters," *Norfolk (NE) Daily News*, 10 July 1993; David Hendee, "Watches, Warnings Continue in Nebraska," *Omaha (NE) World-Herald*, 12 July 1993; Laura Erbst, "Raging Creek Takes Out Stanton County Bridge" *Norfolk (NE) Daily News*, 12 July 1993; Greg Wees, "Erosion Threatens Bridges, Roads," *Norfolk (NE) Daily News*, 22 July 1993; Mary Pat Finn-Hoag, "Flood Control Projects Working: SCS Official," *Norfolk (NE) Daily News*, 17 July 1993; Jerry Guenther, "Norfolk's Water Problems 'Minor,'" *Norfolk (NE) Daily News*, 31 July 1993; Jerry Guenther, "Channel Saves City from Flooding," *Norfolk (NE) Daily News*, 31 July 1993.
58. Fred Thomas, "Work Set to Start on Project to Control Elkhorn Flooding," *Omaha (NE) World-Herald*, 14 January 1994.
59. Helms, "Getting to the Roots."

Conclusion

1. U.S. Army Corps of Engineers, Omaha District, *Review Report on Improvement for Flood Control and Allied Purposes, Elkhorn River and its Tributaries, Nebraska, Appendix VIII, Transcript of Public Hearings* (Omaha: War Department, 1947), 149, 280.
2. West, *The Contested Plains*, 233.
3. Kelly Bartnick, "Rising Elkhorn River Prompts Water Rescue," accessed 3 April 2015, http://www.scrippsmedia.com/kmtv/news/Rising-Elkhorn-iver-Prompts-Water-Rescue-263234581.html.
4. "Floodway Renovation Includes Clearing Land of Buildings, Appliances, and Tires," *Norfolk (NE) Daily News*, 5 December 2003; Jerry Guenther, "New Maps Showing Flood Plains Anger Land Owners," *Norfolk (NE) Daily News*, 23 March 2004; "Could Farmers Be of More Help?" *Norfolk (NE) Daily News*, 23 July 2007.
5. Chris Olson, "Waterloo Project Dammed by Flooding Concerns," *Omaha (NE) World-Herald*, 10 October 2005; Chris Olson, "Waterloo Forging Pact on Liability," *Omaha (NE) World-Herald*, 11 December 2005; Chris Olson, "Neighborhood Achieves Deal over Water Liability," *Omaha (NE) World-Herald*, 14 December 2005. For a discussion of low-income housing on land behind levees, see Steinberg, *Acts of God*, 97–115.
6. Kristin Zagurski, "Officials Urge Rivers' Neighbors to Prepare for Possible Flooding," *Omaha (NE) World-Herald*, 23 February 2007.
7. "Flood Fears Rise Along Eastern Rivers," *Lincoln (NE) Journal Star*, 7 March 2010.
8. Kevin Cole, "High Water Closes Elkhorn River Rec Areas," *Omaha (NE) World-*

Herald, 10 June 2010; Andrew J. Nelson and Jason Kuiper, "Floodwaters Close Roads Big and Small," *Omaha (NE) World-Herald*, 11 June 2010.

9. Jerry Guenther and Kent Warneke, "Heavy Rainfall Causes River, Creek Levels to Rise," *Norfolk (NE) Daily News*, 12 June 2010; Nancy Gaarder and Kevin Cole, "Water Empties Towns, Erases Roads—Gov. Heineman Raises the Alarm about Flooding So Far and Yet to Come," *Omaha (NE) World-Herald*, 14 June 2010; David Hendee, "Water Fight," *Omaha (NE) World-Herald*, 20 June 2010.
10. Nancy Gaarder and David Hendee, "Railroad Worker Still Missing in Elkhorn River," *Omaha (NE) World-Herald*, 16 June 2010; David Hendee, "Water Fight," *Omaha (NE) World-Herald*, 20 June 2010.
11. "The Legend of the Swinging Bridge," sign in Riverside Park, photograph in author's collection; LuAnn Schindler, "Neligh Historic Site Begins Rebuilding," *Norfolk (NE) Daily News*, 14 December 2010.
12. LuAnn Schindler, "Neligh Historic Site Begins Rebuilding," *Norfolk (NE) Daily News*, 14 December 2010; quotation in Sheryl Schmeckpeper, "River, Mill Important to Town's History," *Norfolk (NE) Daily News*, 31 August 2013.
13. "Norfolk's 100–year flood," *Norfolk (NE) Daily News*, 17 June 2010; streamflow data retrieved from United States Geological Survey, National Water Information System, "USGS 06799000 Elkhorn River at Norfolk, Nebr.," http://nwis.waterdata.usgs.gov/ne/nwis/uv/?cb_00060=on&cb_00065=on&format=gif_default&site_no=06799000&period=&begin_date=2010-01-01&end_date=2015-08-19 (accessed 19 August 2015).
14. Joe Duggan, "1 Missing after Rail Bridge Collapses," *Lincoln (NE) Journal Star*, 16 June 2010; Nancy Gaarder and David Hendee, "Railroad Worker Still Missing in Elkhorn River," *Omaha (NE) World-Herald*, 16 June 2010; Joe Duggan, "Dealing with the Aftermath—Harrowing Rescue Saves One, But Officials Are Haunted by Missing Worker," *Lincoln (NE) Journal Star*, 17 June 2010.
15. "Body of Missing Railroad Worker Found," *Norfolk (NE) Daily News*, 18 June 2010; "Norfolk Flood Historic," *Lincoln (NE) Journal Star*, 18 June 2010.
16. Duggan, "Dealing with the Aftermath."
17. Kathryn Harris and Courtney Pitts, "Flood Waters Impact Area Industries," *Norfolk (NE) Daily News*, 16 June 2010; Trisha Schulz, "Businesses Assess, Clean Up Flood Damage," *Norfolk (NE) Daily News*, 18 June 2010; Courtney Pitts, "After the Flood," *Omaha (NE) World-Herald*, 26 June 2010; Sheryl Schmeckpeper and Kathryn Harris, "Flood's Impact Felt Long After Water Recedes," *Norfolk (NE) Daily News*, 15 June 2011.
18. Paul Hammel, "Damage Stirs Memories, Money Worries," *Omaha (NE) World-Herald*, 18 June 2010; Joe Duggan, "Elkhorn River Bucks the Cowboy Trail," *Lincoln (NE) Journal Star*, 30 August 2010.

19. "Assistance for Flood Victims Announced," *Norfolk (NE) Daily News*, 16 June 2010; Alissa Skelton, "State's Risks for More Flooding, Severe Weather Not Over," *Lincoln (NE) Journal Star*, 20 June 2010; Courtney Pitts and Trisha Schulz, "Trucks to Replace Rail Transportation," *Norfolk (NE) Daily News*, 22 June 2010.
20. Courtney Pitts and Trisha Schulz, "Trucks to Replace Rail Transportation," *Norfolk (NE) Daily News*, 22 June 2010; Kathryn Harris, "Johanns Tours Flood-Damaged Areas, Promises Push for Results," *Norfolk (NE) Daily News*, 6 July 2010; Greg Wees, "Corps, City Mull Bridge Issue," *Norfolk (NE) Daily News*, 27 July 2010.
21. Delores Meister, "West Point Recovering from Flood," *Norfolk (NE) Daily News*, 1 July 2010.
22. Algis J. Laukaitis, "Almost-finished Lake Wanahoo Dam Prevents Flood," *Lincoln (NE) Journal Star*, 28 June 2010; Sarah Reinecke, "Red Cross Reliables Always on Front Lines," *Omaha (NE) World-Herald*, 28 June 2010.
23. Duggan, "Elkhorn River Bucks the Cowboy Trail."
24. "Officials Prepare Plan to Fix Elkhorn River Channel Issue," *Lincoln (NE) Journal Star*, 14 September 2010.
25. Kevin Cole, "Residents of King Lake Make Gains against Flood," *Omaha (NE) World-Herald*, 20 September 2010.
26. Greg Wees, "Railroad Recovered from the Flood," *Norfolk (NE) Daily News*, 14 June 2011.
27. Trisha Schulz, "Companies Give Thanks for Speedy Bridge Recovery," *Norfolk (NE) Daily News*, 12 October 2010.
28. Asha Anchan, "Rescue Workers Recall Last Year's Flood," *Norfolk (NE) Daily News*, 13 June 2011.
29. Trisha Schulz, "Flood Victim's Widow Reveals She's Expecting," *Omaha (NE) World-Herald*, 26 August 2010; Trisha Schulz, "A Year Later, Family Still Coping with Loss," *Norfolk (NE) Daily News*, 13 June 2011; "Family of Man Killed in Bridge Collapse Sues," *Lincoln (NE) Journal Star*, 21 May 2011; Trisha Schulz, "Wrongful Death Suit Settled," *Norfolk (NE) Daily News*, 1 August 2013.
30. "Stubborn Storm Wreaks Havoc," *Denver (CO) Post*, accessed 12 October 2015, https://html2-f.scribdassets.com/7nypffo43k2wvw3d/images/1–8a60cc3aaf.jpg.
31. Tina Burgess, "Colorado Floods: Colorado Road Closures and Latest Colorado Flood Chaos Updates," *San Diego (CA) Examiner*, 16 September 2013.
32. Mark Jaffe, "Colorado Floods: 1,900 Oil and Gas Wells Shut as Crews Check Damage," *Denver (CO) Post*, 17 September 2013, accessed 12 October 2015, http://www.denverpost.com/breakingnews/ci_24116404/oil-field-flood-tally-1-900-wells-shut.
33. Ivan Moreno, "Colo. Immigrants Face Uncertainty after Floods," *Denver (CO) Post*, 27 September 2013, accessed 12 October 2015, http://www.denverpost.com/news/ci_24189998/immigrants-face-uncertainty-after-colo-floods.

34. Ivan Moreno and Manuel Valdes, "For Colorado Farms, Floods Could Bring Recharged Soil," *Denver (CO) Post*, 17 September 2013, accessed 12 October 2015, http://www.denverpost.com/2013coloradofloods/ci_24119125/colorado-farms-floods-could-bring-recharged-soil.
35. Algis J. Laukaitis, "Rainwater Basin, Platte River Could See Benefits from Floodwaters," *Lincoln (NE) Journal Star*, 30 September 2013; Heather Johnson, "Finding the Silver Lining in the Flood," *North Platte (NE) Telegraph*, 24 September 2013.
36. Nancy Gaarder, "Unprecedented Flooding Feared; Nebraskans Take Precautions," *Omaha (NE) World-Herald*, 17 September 2013.
37. Tyler Ellyson, "More Water, No Flooding Expected in Area," *Columbus (NE) Telegram*, 18 September 2013.
38. Nancy Gaarder, "Towns Scurry as Water Rushes In—South Platte Communities Dealing with Colorado Runoff," *Omaha (NE) World-Herald*, 18 September 2013.
39. Grant Schulte, "Nebraska Officials Continue Work as Flood Worries Ease along Platte," *Lincoln (NE) Journal Star*, 21 September 2013, accessed 13 October 2015, http://journalstar.com/news/state-and-regional/nebraska/nebraska-officials-continue-work-as-flood-worries-ease-along-platte/article_43c78ee2-6cf0-56db-89c5-1c056464d88a.html; Nancy Gaarder, "Floodwaters Reach New Heights in Three Towns—Early Preparation Helps Prevent Much Damage in Nebraska Communities," *Omaha (NE) World-Herald*, 25 September 2013.
40. "Flood Crest Has Passed Kearney But Platte Still Running High," *Kearney (NE) Hub*, 30 September 2013; Nancy Gaarder, "Nebraska Floodwaters Sampled for Contaminants," *Omaha (NE) World-Herald*, 21 September 2013; Tyler Ellyson, "Platte River Could Be Unsafe for Days, Weeks," *Columbus (NE) Telegram*, 29 September 2013.
41. Staff Writer, "The Mighty South Platte River Rises Again," *North Platte (NE) Telegraph*, 27 September 2013, accessed 12 October 2015, http://www.nptelegraph.com/sports/outdoors/the-mighty-south-platte-river-rises-again/article_baa17bca-8c73-5f1a-b307-62821c714312.html#facebook-comments.
42. Staff Writer, "The Mighty South Platte River Rises Again."
43. Donald Worster, "The American West in the Age of Vulnerability," *Western Historical Quarterly* 45 (Spring 2014): 7–8.
44. Ibid., 9.
45. Ibid.
46. Ibid., 11.
47. Ibid., 12.
48. Steinberg, *Acts of God*, xv–xvi.
49. Worster, "The American West in the Age of Vulnerability," 13–14.
50. Al J. Laukaitis, "Platte Ice Raises Fear of Flooding," *Lincoln (NE) Journal Star*, 6 January 2001.
51. Klein and Zellmer, *Mississippi River Tragedies*, 203.

52. Alan Lightman, "Our Lonely Home in Nature," *New York Times*, 2 May 2014, accessed 3 May 2014, http://www.nytimes.com/2014/05/03/opnion/our-lonely-home-in-nature.html?_r=0.
53. Klein and Zellmer, *Mississippi River Tragedies*, 203.
54. Bob Dylan, "Buckets of Rain," *Blood On the Tracks*, LP, Columbia PC 33235, 1975.

SOURCES

Primary

Newspapers

Bangor (ME) Daily Whig and Courier, 1855. Available online through various databases; on microfilm at the Library of Congress; original copies available at Bangor Historical Society, Bangor, ME.

Boston (MA) Daily Advertiser, 1873. Available online through various databases; on microfilm at the Library of Congress; original issues available at American Antiquarian Society, Newspaper Project, Worcester, MA.

Columbus (NE) Telegram.

Cuming County (NE) Advertiser, 1891.

Cuming County (NE) Democrat, 1947.

Daily Arkansas Gazette (Little Rock), 1875. Available online through various databases; originals held at Arkansas Historical Commission, Little Rock, AR.

Daily Evening Bulletin (SF), 1873. Available online through various databases; originals held at American Antiquarian Society, Newspaper Project, Worcester, MA, and San Francisco Public Library, San Francisco, CA.

Daily Rocky Mountain News (Denver, CO), 1873. Available online through various databases; microfilm held at Denver Public Library, Denver, CO.

Fremont (NE) Weekly Herald, 1881, 1940, 1944.

Frontier Guardian (Kanesville [i.e., Council Bluffs], IA. Available in digital form at The Church of Jesus Christ of Latter-Day Saints Church History Library, Salt Lake City, Utah. Available at https://eadview.lds.org/findingaid/M205.1%20F935%201849-1853/.

Hooper (NE) Sentinel.

Kearney (NE) Hub.

Lincoln (NE) Journal Star, 2010.

Neligh (NE) Leader, 1888, 1944.

Norfolk (NE) Daily News, 1888.

North Platte (NE) Telegraph.

Omaha (NE) World-Herald, 1984, 1990, 1993, 1994, 2005, 2007, 2010, 2013.

St. Louis (MO) Globe-Democrat, 1877. This paper's archives can be found in the St. Louis Mercantile Library, University of Missouri-St. Louis.

Stanton (NE) Register, 1944.

West Point (NE) Republican, 1873, 1920, 1940.

West Point (NE) Republican and Cuming County Advertiser, 1912.

Note: Nebraska newspapers can be found in the archives of the Nebraska State Historical Society in Lincoln.

Other

Hansen, Esther Kolterman. *Along Pioneer Trails in Pierce County, Nebraska*. N.p., 1940. Available in local history library, Nebraska State Historical Society.

Harris, Sarah Hollister. *An Unwritten Chapter of Salt Lake, 1851–1901*. New York: N.p., 1901. Available online at https://history.lds.org/overlandtravels/trailExcerptMulti?lang=eng&companyId=58&sourceId=9288, accessed 20 May 2015.

Jones, J.W.C. "The Elkhorn Valley; its climate, soil, surface, water, timber grasses, products, people, and industrial resources; or, A month's talk, walk, and drive, in the counties of Cuming, Stanton, Madison, Antelope, Wayne and Pierce, Nebraska." West Point, NE: Republican Job Office, 1880.

Nebraska Department of Natural Resources. "Small Watersheds Flood Control Fund." Available at http://dnr.nebraska.gov/small-watersheds-flood-control-fund, accessed 12 August 2015.

———, Floodplain and Dam Safety Division. *Flood Hazard Mitigation Plan, State of Nebraska*. Lincoln: Nebraska Department of Natural Resources, 2013.

Nebraska Natural Resources Commission. *Cuming County, Nebraska*. Vol. 1, *Flood Plain Study: Elkhorn River*. Lincoln: Nebraska Natural Resources Commission, 1987.

———, Water Quality Planning Staff. *Elkhorn River Basin Water Quality Management Plan*. Lincoln: Nebraska Natural Resources Commission, 1975.

Nebraska Soil and Water Conservation Commission. *Flood Prevention and Damage Reduction*. State Water Plan, Publication Number 401, Revised Printing. Lincoln: The Commission, 1968.

Nebraska State Planning Board. *Water Resources of Nebraska*. Lincoln: N.p., 1941.

U.S. Army Corps of Engineers, Omaha District. *Detailed Project Report: Flood Protection Project at Scribner, Pebble Creek at Scribner, Nebraska*. Omaha: U.S. Army Corps of Engineers, 1985.

———. *Flood Protection Project: Pebble Creek and Elkhorn River at Scibner, Nebraska; Supplement to Detailed Project Report*. Omaha: U.S. Army Corps of Engineers, 1991.

———. *Project Maps, Omaha District: Status of River and Harbor and Flood Control Projects*. Omaha: Corps of Engineers, 1983.

———. *Review Report on Improvement for Flood Control and Allied Purposes, Elkhorn River and Its Tributaries, Nebraska*. Omaha: War Department, 1947.

———. *Review Report on Improvement for Flood Control and Allied Purposes, Elkhorn River and Its Tributaries, Nebraska, Appendix VIII, Transcript of Public Hearings*. Omaha: War Department, 1947.

———. *Supplement to Detailed Project Report: Flood Protection Project, Pebble Creek and Elkhorn River at Scribner, Nebraska*. Omaha: U.S. Army Corps of Engineers, 1991.

U.S. Engineer Office. *Interim Report on Flood Control for the City of Norfolk, Elkhorn River Basin, Nebraska*. Omaha: U.S. Engineer Office, 1946.

U.S. Geological Survey, National Water Information System. "USGS 06799000 Elkhorn River at Norfolk, Nebr." Available at http://nwis.waterdata.usgs.gov/ne/nwis/uv/?cb_00060=on&cb_00065=on&format=gif_default&site_no=06799000&period=&begin_date=2010-01-01&end_date=2015-08-19, accessed 19 August 2015.

Wilhelm, Paul, Duke of Württemberg. Trans. W. Robert Nitske, ed. Savoie Lottinville. *Travels in North America, 1822–1824*. Norman: University of Oklahoma Press, 1973.

Secondary

Amundson, Michael A. "Portrait of a Small Town: The Photographic Diary of Neligh's Emanuel Wolfe." *Nebraska History* 79 (Winter 1998): 150–151.

Archer, Kenna Lang. *Unruly Waters: A Social and Environmental History of the Brazos River*. Albuquerque: University of New Mexico Press, 2015.

Bailey, Lloyd R. *Noah: The Person and the Story in History and Tradition*. Columbia: University of South Carolina Press, 1989.

Bentall, Ray. *Water Supplies and the Land: The Elkhorn River Basin of Nebraska*. Lincoln: University of Nebraska, 1971.

Blakeslee, Donald J. "Assessing the Central Plains Tradition in Eastern Nebraska: Content and Outcome." In *The Central Plains Tradition: Internal Development & External Relationships*, edited by Donald J. Blakeslee. Iowa City: Office of the State Archaeologist, The University of Iowa, 1978.

———, and Warren W. Caldwell. *The Nebraska Phase: An Appraisal*. Lincoln, NE: J & L Reprint Company, 1979.

Bleed, Ann, and Charles Flowerday, eds. *An Atlas of the Sandhills*. 2nd ed. Lincoln: Conservation and Survey Division, Institute of Agriculture and Natural Resources, and University of Nebraska-Lincoln, 1990.

Bleed, Ann, Jim Barr, Dayle Williamson, and Gayle Starr. "The Creation of Nebraska's Natural Resource Districts." *Nebraska History* 97 (Winter 2016): 199–214.

Bomer, Norm. *Sons of the River: A Nebraska Memoir*. Moscow, ID: Canon Press, 2000.

Buecker, Thomas R. *Water Powered Flour Mills in Nebraska*. Lincoln: Nebraska State Historical Society, 1983.

Carter, John E. "Niobrara, Nebraska: The Town Too Tough to Stay Put!" *Nebraska History* 72 (Fall 1991): 144–149.

Ciocc, Mark. *The Rhine: An Eco-Biography, 1815–2000*. Seattle and London: University of Washington Press, 2002.

Colten, Craig E. *An Unnatural Metropolis: Wresting New Orleans from Nature*. Baton Rouge: Louisiana State University Press, 2005.

Dearen, Patrick. *Bitter Waters: The Struggles of the Pecos River*. Norman: University of Oklahoma Press, 2016.

DeBell, Ella J. "The Bruner and Neligh Families." *Nebraska History* 16 (April–June 1935): 82–92.

deBuys, William. *A Great Aridness: Climate Change and the Future of the American Southwest*. New York: Oxford University Press, 2011.

Dylan, Bob. "Buckets of Rain." *Blood On the Tracks*. LP, Columbia PC 33235, 1975.

Fenn, Elizabeth A. *Encounters at the Heart of the World: A History of the Mandan People*. New York: Hill and Wang, 2014.

Fitzpatrick, Lillian L. *Nebraska Place-Names*. Edited by G. Thomas Fairclough. Lincoln: University of Nebraska Press, 1960.

Fletcher, Alice, and Francis La Flesche. *The Omaha Tribe*, 2 vols. Washington, DC: Government Printing Office, 1911. Reprint, Lincoln: University of Nebraska Press, 1972.

Gibson, Arrell Morgan. *The American Indian: Prehistory to Present*. Lexington, MA: D. C. Heath and Company, 1980.

Grier, Bob. "The Elkhorn." *Nebraskaland*. Special edition, *Nebraska Rivers* 61 (January–February 1983).

Haas, Dan. "Wild and Scenic Rivers." In *Encyclopedia of the Great Plains*, edited by David J. Wishart. Lincoln: University of Nebraska Press, 2004.

Harvey, Amy E. *Oneota Culture in Northwestern Iowa*. Iowa City: University of Iowa, 1979.

Helms, Douglas. "Getting to the Roots." In *People Protecting Their Land: Proceedings, Volume 1*, 7th ISCO Conference. Sydney, Australia: International Soil Conservation Organization, 1992. Available at http://www.nrcs.usda.gov/wps/portal/nrcs/detail/national/about/history/?&cid=nrcs143_021394, accessed 24 September 2015.

Holtz, Milton E. "Early Settlement and Public Land Disposal in the Elkhorn River Valley, Cuming County, Nebraska Territory." *Nebraska History* 52 (Summer 1971): 113–132.

Hoover, Herbert T., John Rau, and Leonard Bruguier. "Gorging Ice and Flooding Rivers: Springtime Devastation in South Dakota." *South Dakota History* 17 (Summer–Fall 1987): 181–201.

Hoyt, William G., and Walter B. Langbein. *Floods*. Princeton, NJ: Princeton University Press, 1955.

Hundley, Jr., Norris. *The Great Thirst: Californians and Water, 1770s–1990s*. Berkeley: University of California Press, 1992.

———. "Water and the West in Historical Imagination." *Western Historical Quarterly* 27 (Spring 1996): 5–31.

———. "Water and the West in Historical Imagination: Part Two—A Decade Later." *The Historian* 66 (Fall 2004): 455–490.

Hurt, R. Douglas. *The Big Empty: The Great Plains in the Twentieth Century*. Tucson: University of Arizona Press, 2011.

———. *The Dust Bowl: An Agricultural and Social History*. Chicago: Nelson-Hall, 1981.

Ingram, B. Lynn, and Frances Malamud-Roam. *The West without Water: What Past Floods, Droughts, and Other Climatic Clues Tell Us about Tomorrow*. Berkeley: University of California Press, 2013.

Isenberg, Andrew C. *The Destruction of the Bison: An Environmental History, 1750–1920*. Cambridge: Cambridge University Press, 2000.

Jablow, Joseph. *Ethnohistory of the Ponca*. New York: Garland Publishing, 1974.

Jackovich, Karen G. "Carson's Homecoming." *People Weekly* 16 (December 7, 1981). Available at http://people.com/archive/cover-story-carsons-homecoming-vol-16-no-23/, accessed 4 November 2014.

Jenkins, Hazel M. *A History of Nebraska's Natural Resource Districts*. Lincoln: Nebraska Department of Natural Resources, 1975; revised 2009. Available at http://dnr.ne.gov/nebraska-s-natural-resources-districts-history-updated-july-2009, accessed 29 September 2015.

Jordan, Paul R. "Data for Selected Rivers." In *Encyclopedia of the Great Plains*, edited by David J. Wishart, 861–862. Lincoln: University of Nebraska Press, 2004.

Karskens, Grace. "Floods and Flood-mindedness in Early Colonial Australia." *Environmental History* 21 (April 2016): 315–342.

———. *People of the River*. Sydney: Allen & Unwin. Forthcoming.

Kelman, Ari. *A River and Its City: The Nature of Landscape in New Orleans*. Berkeley: University of California Press, 2003.

Kerstetter, Todd M. "High Water: Flooding in the Elkhorn Basin, 1823–1940." *Nebraska History* 76 (Winter 1995): 176–187.

———. "'The Worst Floods in History': The Federal Government and the Floods of 1944 in the Elkhorn River Basin." *Great Plains Quarterly* 21 (Summer 2001): 179–192.

———. "Up the River: Flooding and Flood Control in Nebraska's Elkhorn River Basin through the Floods of 1944." Master's thesis, University of Nebraska, 1992.

Kline, Christine A., and Sandra B. Zellmer. *Mississippi River Tragedies: A Century of Unnatural Disaster*. New York: New York University Press, 2014.

Kreiter Directory Co's. Norfolk City and Madison County Directory, 1944–45. Vol. XV. Rochester, MN, and Norfolk, NE: Kreiter Directory Co., 1944.

Lakota Winter Counts: An Online Exhibit. Smithsonian Institution, National Anthropological Archives, and National Museum of the American Indian [2002]. Available at http://wintercounts.si.edu/flashindex.html, accessed 26 January 2016.

Langston, Nancy. *Where Land and Water Meet: A Western Landscape Transformed.* Seattle and London: University of Washington Press, 2003.

Lathrop, Joseph, D.D. "Moral Reflections on Floods." Sermon preached at the First Church in West Springfield, MA, 22 January 1807. Springfield: H. Brewer, Printer, 1807.

Lawson, Michael L. *Dammed Indians: The Pick-Sloan Plan and the Missouri River Sioux, 1944–1980*. Norman: University of Oklahoma Press, 1982.

Leach, A.J. *A History of Antelope County, Nebraska, From Its First Settlement in 1868 to the Close of the Year 1883*. Chicago: The Lakeside Press, 1909.

Leopold, Luna B. *Water: A Primer*. San Francisco: W. H. Freeman and Company, 1974.

Lightman, Alan. "Our Lonely Home in Nature." *New York Times*, 2 May 2014. Available at http://www.nytimes.com/2014/05/03/opnion/our-lonely-home-in-nature.html?_r=0, accessed 3 May 2014.

Lower Platte North Natural Resource District. "About Us." Available at http://www.lpnnrd.org/about/history.html, accessed 12 August 2015.

Ludwickson, John. "Central Plains Tradition Settlements in the Loup River Basin: The Loup River Phase." In *The Central Plains Tradition: Internal Development & External Relationships*, edited by Donald J. Blakeslee. Iowa City: Office of the State Archaeologist, The University of Iowa, 1978.

Malin, James C. "Ecology and History." In James C. Malin, *History and Ecology: Studies of the Grassland*, edited by Robert P. Swieringa. Lincoln: University of Nebraska Press, 1984.

Martin, Kent E. *Neligh Mill State Historic Site, Neligh, Nebraska, Nebraska State Historical Society: Self-Guided Tour*. Lincoln: Nebraska State Historical Society, 1997.

Mattison, Ray H. "The Flood of 1881." *North Dakota History* 34 (Summer 1967): 225–241.

Miller, John. "From the Great Plains to L.A.: The Intersecting Paths of Lawrence Welk and Johnny Carson." *Virginia Quarterly Review* 79 (Spring 2003): 265–279.

Morris, Christopher. *The Big Muddy: An Environmental History of the Mississippi and Its Peoples from Hernando de Soto to Hurricane Katrina*. New York: Oxford University Press, 2012.

Moulton, Gary E., ed. *August 30, 1803–August 24, 1804*. Vol. 2 of *The Journals of the Lewis and Clark Expedition*. Lincoln: University of Nebraska Press, 1986.

Nebraska Association of Resource Districts. "About NRDs." Available at http://nrdnet.org/about-nrds.php, accessed 12 August 2015.

———. "Find Your NRD." Available at http://nrdnet.org/find-your-nrd.php, accessed 12 August 2015.

Orsi, Jared. *Hazardous Metropolis: Flooding an Urban Ecology in Los Angeles*. Berkeley: University of California Press, 2004.

Pisani, Donald J. *From the Family Farm to Agribusiness: The Irrigation Crusade in California and the West, 1850–1931*. Berkeley: University of California Press, 1984.

Ridgeway, Marian E. *The Missouri Basin's Pick-Sloan Plan: A Case Study in Congressional Policy Determination*. Illinois Studies in the Social Sciences, vol. 35. Urbana: University of Illinois Press, 1955.

Sanborn Map Company. *Insurance Map of West Point, Nebraska*. New York: Sanborn Map Company, 1886, 1889, 1897, 1904, 1909, 1921.

Schmitt, Frank. "A Pioneer Nebraska Community." *Nebraska History* 18 (April–June 1937): 77–90.

Schneiders, Robert Kelley. *Unruly River: Two Centuries of Change along the Missouri*. Lawrence: University Press of Kansas, 1999.

Scoville, C. H., comp. *History of the Elkhorn Valley, Nebraska: An Album of History and Biography Containing a Descriptive Political, Statistical, Financial, Commercial, Agricultural, Educational, Religious and Biographical History and Reminiscences*. Chicago and Omaha: National Publishing Company, 1892. Reproduced by Unigraphic, Inc., Evansville, IN, 1977.

Sherow, James E. "Water." In *Encyclopedia of the Great Plains*, edited by David J. Wishart. Lincoln: University of Nebraska Press, 2004.

———. *Watering the Valley: Development along the High Plains Arkansas River, 1870–1950*. Lawrence: University Press of Kansas, 1990.

Smith, G. Hubert. "Notes on Omaha Ethhnohistory, 1763–1820." *Plains Anthropologist* 18 (November 1973): 257–264.

Steinberg, Ted. *Acts of God: The Unnatural History of Natural Disaster in America*. New York: Oxford University Press, 2000.

Stubbs, Stacey. "'Gentle River Goes Mad': The Republican River Flood of 1935 and Its New Deal Legacy." *Nebraska History* 97 (Spring 2016): 2–15.

Webb, Walter Prescott. *The Great Plains*. Boston: Ginn, 1931. Reprint, Lincoln: University of Nebraska Press, 1981.

Wedel, Waldo R. *Prehistoric Man on the Great Plains*. Norman: University of Oklahoma Press, 1961).

West, Elliott. *The Contested Plains: Indians, Goldseekers, and the Rush to Colorado*. Lawrence: University Press of Kansas, 1998.

———. *The Way to the West: Essays on the Central Plains*. Albuquerque: University of New Mexico Press, 1995.

White, Richard. *The Organic Machine*. New York: Hill and Wang, 1995.

———. *The Roots of Dependency: Subsistence, Environment, and Social Change among the Choctaws, Pawnees, and Navajos*. Lincoln: University of Nebraska Press, 1983.

Wishart, David J. "The Great Plains Region." In *Encyclopedia of the Great Plains*, edited by Wishart, viii–xviii. Lincoln: University of Nebraska Press, 2004.

——, ed. "Water." In *Encyclopedia of the Great Plains*, edited by Wishart, 844–867. Lincoln: University of Nebraska Press, 2004.

Worster, David. *The Dust Bowl: The Southern Plains in the 1930s*, 25th Anniversary Edition. New York: Oxford, 2004.

Worster, Donald. "The American West in the Age of Vulnerability." *Western Historical Quarterly* 45 (Spring 2014): 5–16.

——. *Dust Bowl: The Southern Plains in the 1930s*. New York: Oxford University Press, 1979.

——. *Rivers of Empire: Water, Aridity, and the Growth of the American West*. New York: Oxford, University Press, 1985.

INDEX

Page numbers in *italics* represent figures.

aboriginal peoples, 28
Achziger, Lyle, 151
Adams, Virginia, 147
Adkins, B.E., 106
Agricultural Adjustment Administration, 88–89, 106
American Red Cross, 87–88, 102–3, 106, 131–34, 147
Anheuser-Busch, 146
Antelope County, Nebraska, 19, 45, 57, 90, 106, 146
    saw mills, 47–48. *See also* Elkhorn River, milling
Anthropocene epoch, xv, 155, 159
Arawaks, 27
Archaic tradition, 36–37
Arkfeld Manufacturing and Distributing, 145
Arlington, Nebraska, 19, 87, 97, 99, 101
Association of State Floodplain Managers, 35
Aswan Dam, 74
Atkinson, Nebraska, 19, 40, 91
Audubon Society, 152

Babbitt, Almon W., 39
Bancroft, Nebraska, 73
Bancroft mill, 75
Battle Creek, Nebraska, 19, 94, 109, 111, 116–17, 119, 123, 140, 146
Beemer, 19, 80, 86, 96–97, 101, 123, 146
Bell Creek, 99, 108
Bell, Robert M., 126
Big Springs, Nebraska, 153
Bismarck, Dakota Territory, 31
Blood on the Tracks, 3–4
Blue River, xiii
Bomer, Norm, 9–10
Boone County, Nebraska, 90
Brazos River, 11
Breyer, A.W., 94
Brown County, Nebraska, 19
Bruett, Cameron, 151
Brulé Lakota Indians, 38
Brule, Nebraska, 153
Bruner, Uriah, 44–46
"Buckets of Rain," 4, 161
Buffalo Creek, 109, 111, 114–15, 117
Burt County, Nebraska, 96
Butterfly Creek, 97

California Aquarium car, 50–52
Canadian Heritage Rivers System, 12
Carlson, G.L., 92
Carleton, Ron, 151

Carson, Dick, 10
Carson, Johnny, 10
Cedar County, Nebraska, 90
Cedar Creek, 73, 108
Central Nebraska Public Power and Irrigation District, 152
Central Plains tradition, Nebraska Phase, 37
Chamberlain, Roy, 97
Cheyenne Indians, 38
Chicago and North Western Railroad, 51, 86–88, 101, 118–19, 145
Chicago, St. Paul, Minneapolis, and Omaha Railroad, 65–66, 76
China, 27. *See also* flooding
Church of Jesus Christ of Latter-day Saints, 39
Clarkson, Nebraska, 97–98
Clearwater Creek, 48
Clearwater, Nebraska, 7–8, 19, 71, 143
climate change, 16, 154, 157, 160, 161
Colfax County, Nebraska, 96, 98
Columbus, Nebraska, 94, 153, 155
Cowboy Trail, 145, 147
Crawford, J.C., 44
Crowell, Nebraska, 19, 97
Crutzen, Paul J., 155
Cuming County, Nebraska, 19–20, 55, 72–73, 88, 96, 106
    Board of Supervisors, 103
    fairgrounds, 96

Dakota Indians, 29–30
Dakota Southern Railroad, 31
Dakota Territory, 31
Davis, C.A., 80
Davis, Pete, 94
Denver Basin, 151
Dial Land Development Corporation, 140
Diers, E.C., 107
disaster, meaning of, xiv–xvi, 13–15, 43, 49, 54, 95, 105, 154–56. *See also* natural disaster, unnatural disaster
Dodge County, Nebraska, 19, 73, 87, 96, 104, 142, 147–48
Dodge, Nebraska, 97
Douglas County, Nebraska, 19, 87, 147–48
drainage districts, 67–69, 75, 81, 158
drought, 156
Dry Creek, 19, 81
Duke of Württemberg (Paul Wilhelm), 35–36, 38–39, 74, 157
Dunnigan, Brian, 152
Dust Bowl: The Southern Plains in the 1930s, 115–16
Dylan, Bob, 3, 161

East Fork Creek, 98
East Pierre, Dakota Territory, 31–32
Encyclopedia of the Great Plains, 12
Elkhorn Middle School, 146
Elkhorn, Nebraska, 19, 39
Elkhorn River
    bed characteristics, 22
    compared to other rivers, 10, 17
    cultural interpretations, 6–11
    dams, 45–50
    dialogue with humans, xv–xvi, 34, 36, 43–44, 53, 57, 65, 90, 120, 140, 157
    described, 19
    dredging, 7
    evacuation planning 141
    fatalities related to floods, 50, 52, 76, 80, 101, 133, 144, 149–50
    fish species, 23, 51–52
    floods, 10
    flood of 1823, 35–36
    flood of 1873, 49–52
    flood of 1877, 52

flood of 1881, 53–54
flood of 1888, 54–55
floods of 1891, 55–57
floods of 1897, 65–67
flood of 1912, 69–74, 122
floods of 1917, 75–78
flood of 1920, 79–81
flood of 1940, 86–90
floods of 1944, 5, 16, 90–103
floods of 1947, 118–20
flood of 1960, 121–22
flood of 1962, 122–24
flood of 1966, 124–25
flood of 1971, 130
flood of 1984, 131–32
flood of 1990, 133–34
floods of 1993, 134–36
flood of 2010, 5, 16, 142–50
floodplain development, xiv, xv
flow, 22
Great Plains river, as a 12–13, 17
hydropower, 42–50, 52, 56, 60–65
ice source (commodity), 69–70. *See also* ice jams and flooding
in hydrological system, 155
milling, 22, 44–50, 60–65
mouth, 20
North Fork. See North Fork of Elkhorn River
recreation, 9, 47, 139, 142
sensitivity to flooding, 21
source, 17
South Fork, 19
unnavigable, 22, 40, 43, 46–47
Elkhorn River Basin, xiii, 3–4
booster literature, 40–43
described, 20–21
fauna, 23–24, 57
fertility, 42–43, 57, 157
flood damage estimate for 1940–1965, 128
flora, 22–23, 46
lower basin, 5, 19, 23, 68, 90, 96–97, 107, 133
precipitation, 21, 41–42
upper basin, 9, 17, 22–23, 54–55, 76, 90, 96, 120
settlement of by non-Indians, 15, 44, 59
soil types, 23
transportation corridor, as a, 5, 13, 35–36, 39–40, 57, 157
Elkhorn River Drainage District, 68
Elkhorn Valley Drainage District, 68
Elkhorn Valley Immigration Association, 40–43
Elkhorn Valley Museum, 81
Enlightenment, 154
Euphrates River, 74
Evans, Colorado, 151
Ewing, Nebraska, 9–10, 19, 69, 79
Exon, James, 130

Farm Security Administration, 106
Fauss, Mary Ann, 133
Federal Emergency Management Agency (FEMA), 140, 144
flooding
appearances in global cultures, 27
beneficial aspects, 27
colonial Australia, in, 27–29
China, in, 27
defined, 25. *See also* disaster, natural disaster, unnatural disaster
evil or malicious, 26–27
mechanics of, 15, 24–26
trait of a river, 4
floodplain

defined, 24
development of, 14–15
imperfect landscape, as an, 4
Plains Indians' use of, 35–39
soil fertility, 25, 36
Flood Control Act (1928), 82
Flood Control Act (1936), 82, 104
Flood Disaster Protection Act (1973), 129
Fort Abraham Lincoln, 31
Fremont, Nebraska, 19, 97, 99, 101, 106–7, 111
Fremont Guide and Tribune, 85, 99–100, 107
Fremont Herald, 53
French, Alvin and Gertrude, 132

Genesis, 27
Giles Creek, 109, 111, 116–17
Gilman, Stephen F., 62–64
God, 13–14, 29, 150, 154
Goeken, Andrew, 133–34. *See also* Elkhorn River, fatalities related to floods
Great American Desert, 40
Great Plains
aridity, 11
environment, 3, 7, 11, 35, 68, 122, 154
rivers, 9–10, 12, 14–17, 22, 40–43, 47, 56, 60, 86, 129, 155
Green Island, Nebraska, 31–32
Gretna, Nebraska, 19, 78, 87, 117
Gross, D.L., 104

Hankins, M.J., 107
Hannibal, Missouri, 13–14, 156. *See also* US Army Corps of Engineers, levees
Hannibal Courier-Post, 14
Harris, Sarah, 39
Harvest Trail, 80
Hawkesbury River, 28–29
Henrichs, Donald, 121. *See also* Elkhorn River, fatalities related to floods
Hershey, Nebraska, 153
Hidatsa Indians, 29–30
Hilker, Deb, 142
Holt County, Nebraska, 19–20, 80, 90, 142
Homer, Nebraska, 38
Homes at River Bend, 141
Hooper Inn, 102
Hooper, Nebraska, 19, 53–54, 97, 99, 106, 109, 111, 114–15, 117, 122–23, 130
American Legion chapter, 102
Commercial Club, 102, 124
fire department, 124
Lion's Club, 124
Hopper, Doyle, 145
Howells, Nebraska, 96–98, 102–03
Hoyt, William G., 25
Hugo, Jerr, 145
Humbug Creek, 20, 78, 82, 86, 97, 108
hydropower, xv, 25, 42 56, 68. *See also* Elkhorn River, hydropower; Elkhorn River, milling

IBP plant (West Point), 134
ice jams and flooding, 24–25, 30, 68, 70, 121–25, 130, 134, 156
"Idiot Wind," 4
Industrial Revolution, 155
Inman, Nebraska, 19, 79, 142
Interstate 80, 125, 152
Island Supply, 145

JBS meat packing, 151
Julesburg Basin. *See* Denver Basin

Karskens, Grace, 28
Kaw River, xiii
King Lake subdivision, 147–48
Klein, Christine A., 14–15, 159–60

Knox County, Nebraska, 90
Kremer, Maurice, 129–30

Lakota Indians, 29–30
Langbein, Walter B., 25
Lauritzen, Dean, 134–35
Leffel turbine, 45
Leigh, Nebraska, 97
Leisy, Robert, 101. *See also* Elkhorn River, fatalities related to floods
levee districts, 67–69
levees, 78, 105, 107, 128, 141
  Elkhorn River Basin, xiv–xv, 89, 108–12, 117, 130, 144, 158–59
  Hannibal, Missouri, 14
  Humbug Creek, 82, 86
  Manhattan, KS, xiii
  Mississipi River, 74
  North Fork of Elkhorn River, 54–55
  Pebble Creek, 83
levees-only approach to flood control, 74–75
Lewis and Clark expedition, 39
Lightman, Alan, 159–60
Lincoln, Nebraska, 128, 156
  Fire and Swift Water Rescue Team, 149
Lind, Bud, 117, 132
Lind family, 132
Lindsay, Nebraska, 94
Logan Creek, 20, 47, 73, 75, 87, 108, 112, 117
Logan Mill, 47. *See also* Elkhorn River, milling
Los Angeles, 11
Los Angeles River, 11
Louis Dreyfus Commodities, 144, 149
Love Signs, 145
Lower Elkhorn Natural Resource District, 130, 140, 148
Lower Platte North Natural Resource District, 135, 147

Madison, Nebraska, 72, 94, 135
Madison County, Nebraska, 19, 46, 90–91, 106, 146
Malena family, 98
Mandan, Dakota Territory, 31
Mandan Indians, 29–30
Manhattan, Kansas, xiii–xiv
Maple Creek, 20, 87, 96, 98, 103, 108, 117, 133, 142
Massachusetts Institute of Technology, 159–60
Matthewson, Charles, 47
McKirahan brothers, 44
Meadow Creek, Nebraska, 19
Meadow Grove, Nebraska, 91, 111, 114–15, 123, 146
Melcher, Martin, 97
Meskenthine Creek, 97, 109, 111
Mih-tutta-hang-kusch (Mandan settlement), 30
Mississippi River, 11, 13, 68
  flood of 1927, 82
  flood of 1993, 13–14
Mississippi River Tragedies: A Century of Unnatural Disaster, 14–15
Missouri River, 1, 12–13, 69
  flood of 1711 and 1712, 29
  flood of 1826, 29–30
  flood of 1881, 30–32
  flood of 1903, 67
  flood of 1912, 74
  flood of 1913, 74
  flood of 1943, 32–33
  flood of 1944, 97
  flood of 1993, xv
  flood history, 29–33
Missouri River Basin Project. *See* Pick-Sloan Plan

Mississippi River Commission, 74
Miyoshi, John, 147
Montana, 31
Mormon Trail, 39
Morrison, Frank, 125
Myrtle Beach, South Carolina, 5

National Flood Insurance Act of 1968, 129
National Flood Insurance Program, 136, 159
National Weather Service, 133, 142
natural disaster, xiv, 14–15, 24, 27–28, 38, 43, 74, 89, 150, 152–53, 158
natural resource district. *See* Nebraska Natural Resource Districts
navigation, 12, 16, 60, 67–68, 159
Nebraska Central Railroad, 144–46, 149–50
Nebraska Health Department, 102
Nebraska Department of Natural Resources, 81, 152
Nebraska Emergency Management Agency, 144
Nebraska Flood Plain Management Act, 128
Nebraska Highway 32, 145–46
Nebraska National Guard, 101, 124
Nebraska Natural Resource Districts, 129–31, 159
Nebraska Settlement Association, 44
Nebraska Soil and Water Conservation Commission, 126–27
Nebraska Soil Conservation Service, 135
Nebraska State Highway Department, 87, 125
Nebraska State Historical Society, 144
Nebraska state-level flood control planning, 126–28, 159
Neligh, John D., 44–45, 47
Neligh, Nebraska, 19, 83
    flood of 1881, 54. *See also* Elkhorn River, flood of 1881
    flood of 1888, 55. *See also* Elkhorn River, flood of 1888
    flood of 1912, 76–77. *See also* Elkhorn River, flood of 1912
    flood of 1920, 79. *See also* Elkhorn River, flood of 1920
    flood of 1944, 91. *See also* Elkhorn River, flood of 1944
    flood of 2010, 143–44. *See also* Elkhorn River, flood of 2010
    founding, 40
    photograph of Elkhorn River, 7, 9
    Riverside Park, 76, 98, 120, 143
    Neligh Mill, *47*, *48*, 49, 60–65, *63*, 72, 76–78. *See also*, Elkhorn River, milling
        end of hydropower, 79
    founded, 47–48
    hydroelectric production, 63–64
    relationship with railroad, 56, 62
    State Historic Site, 143
Nelson, Ben, 145
New Deal, 82, 85, 89, 158
New Orleans, 11
New South Wales, Australia, 27–28
New York City Produce Exchange, 32
New York Times, 74
Nickerson, Nebraska, 19, 87, 103, 133, 142
Nile River, 74
Niobrara, Nebraska, 31–32
Niobrara River, 12
Noah, 27
Norfolk Action Committee, 123
Norfolk Area Chamber of Commerce, 145
Norfolk City Auditorium, 146
Norfolk Daily News, 6, 70, 76, 79–80, 87, 94, 102, 121–22, 125, 130–31, 135
Norfolk Drainage Board, 101–2, 118

Norfolk *Journal*, 55
Norfolk, Nebraska, 5, 19, 117
Corporation Gulch project, 126
flood damage estimate for 1940–1965, 128
flood of 1877 52–53. *See also*, Elkhorn River, flood of 1877
flood of 1881, 53. *See also*, Elkhorn River, flood of 1881
flood of 1888, 54–55. *See also*, Elkhorn River, flood of 1888
flood of 1897, 65–67. *See also* Elkhorn River, flood of 1897
flood of 1912, 70–72, 122. *See also* Elkhorn River, flood of 1912
flood of 1917, 76–77. *See also* Elkhorn River, flood of 1917
flood of 1920, 80. *See also* Elkhorn River, flood of 1920
flood of 1944, 91–94. *See also* Elkhorn River, flood of 1944
flood of 1960, 121–22. *See also* Elkhorn River, flood of 1960
flood of 1962, 122–24. *See also* Elkhorn River, flood of 1962
flood of 1966, 125. *See also* Elkhorn River, flood of 1966
flood of 1993, 135. *See also* Elkhorn River, flood of 1993
flood of 2010, 144–46, 148–50. *See also* Elkhorn River, flood of 2010
floodplain management, 140, 146
founding, 39–40
hearing on flood damages, 106–7, 112
home of Johnny Carson, 10
levees, 55, 65–67, 102. *See also*, levees
local protection project recommended by USACE, 110, 116, 123, 159
millworks, 47. *See also* Elkhorn River, milling
Norfolk Avenue, 52–53, 67, 71, 81, 92, 94, 118
postcard, 7
Norfolk Iron and Metal, 144, 149
Norfolk Sanitary District, 82, 140
North Fork of Elkhorn River, 19, 54–55, 66, 82, 109
flood of 1912, 70–72. *See also* Elkhorn River, flood of 1912
flood of 1917, 76–77. *See also* Elkhorn River, flood of 1917
flood of 1940, 86. *See also* Elkhorn River, flood of 1940
flood of 1944, 91, 94–95. *See also* Elkhorn River, flood of 1944
flood of 1962. *See also* Elkhorn River, flood of 1962.
modification suggested by USACE, 110
postcard, 6–7
North Platte Telegraph, 153
Northwestern Bell Telephone, 103
no-till farming, 140
Nucor Steel, 144, 149

Oakdale, Nebraska, 19, 91
Oakland, Nebraska, 47
Ofe, Don "Harv," 143
Ogallala, Nebraska, 153
Oglala Lakota Indians, 38
Oltmans, Steve, 156
Omaha, xiii, 5, 19, 51–52, 54, 69, 105
flood of 1943, 32
Omaha Creek, 38
Omaha Indians, 5, 38, 44, 57
O'Neill, Nebraska, 5, 19, 40, 76, 142

Oneota culture, 37–38
Ordway, Nehemiah G., 32
Oto Indians, 38

Paleo-Indian tradition 36, 57
Papio-Missouri Natural Resource District, 135, 148, 156
Pawnee Indians, 5, 44, 49
Payne and Shanes Pharmacy, 61
Payne Creek Drainage District, 119
Pebble Creek, 20, 82, 96, 99, 108–9, 111, 119
Pecos River, 11
Pender, Nebraska, 86
Perry, Rick, 154–55
Pichler, Gary, 145
Pick, Lewis A., Col., 105
Pick-Sloan Plan, 105
Pickerel Lake, 69
Pierce, Nebraska, 70–71, 76, 82, 91, 116–17, 122, 125
  dam, 110
  postcard, 6–7
Pierce County, Nebraska, 91, 106
Pilger, Nebraska, 19, 78, 86, 97, 121
Pilger Watershed Project, 126, 135
Plains Indians, 15, 29–30
Platte River, xiii, 12–13, 97,
  308 Report, 104–6
  basin, 21
  congressional assessment of flooding, 90
  flood of 1993, 134
  flood of 2013, 150–54
Platte River Recovery Implementation Program, 152
Pleistocene Age, 20
Plum Creek, 96, 98, 108
Progressive Era, 15, 59–60, 68, 85, 89

Ransdell-Humphreys Flood Control Act of 1917, 78
Rawhide Creek, 69, 108, 112, 148
Reconstruction Finance Corporation, 106
Red River, 12
Republican River, 12, 81
Rio Grande Pacific Corp., 150
riprap, 14, 117–18, 146
River and Harbors Act of 1927, 104
Rock County, Nebraska, 19
Rowe Sanctuary, 152

Salvation Army, 133
sandbagging, 24, 54, 76–77, 91, 122–23, 133–34, 142–43, 153
Sandhills, 17, 20–22
Sanne, Leonard, 115
Sarpy County, Nebraska, 19, 135
Scholl, Jeffery, 144, 149–50, 160. *See also* Elkhorn River, fatalities related to floods
Scholl, Suzie, 144, 149–50
Schuyler, Nebraska, 56, 62
Scott, Wardner G., 101, 107
Scribner, Nebraska, 19, 88, 96–97, 99, 101, 109, 114, 123, 128, 147–48
settler colonialism, 13, 15, 27–29, 49, 51, 54
  legacy of, 160–61
Shell Creek, 56, 62
"Shelter from the Storm," 4
Sioux City and Pacific Railroad, 40–41
Skalowski, Ben, 94
Sloan, W.G., 105
Small Watersheds Flood Control Fund, 126
Smith, Earl, 125
snow, 11, 17, 21, 23–26, 30–32, 49, 52, 65, 68–70, 73, 75, 79, 118, 122, 124, 130, 134, 141–42, 156

Snyder, Nebraska, 97, 103
Sodomka family, 98
Sons of the River: A Nebraska Memoir, 9–10
South Fork of the Elkhorn River, 81
Spatz family, 98
Spring Creek, 110
Stafford Disaster Relief and Emergency Assistance Amendments (1988), 129
Stanton County, Nebraska, 19–20, 57, 95, 97, 106, 120
Stanton, Nebraska, 19, 69, 86–87, 97, 142
Stanton Soil Conservation District, 107, 119
Stefan, Karl, 108
Steinberg, Ted, 13–15
Stoermer, Eugene F., 155
Stone, Livingstone, 50–51
Sugar City Cereal Mill, 72
Swanson, Steve, 147

Taddicken, Bill, 152
Ta-Ha-Zouka Park, 7–8, 121, 124
Taylor Creek, 117
Ten Eyck, Debbie, 132
Theisen Construction, Inc., 149
thunderstorms, 22, 24–26, 29, 38, 42, 55, 68, 96, 125, 142
Tilden, Nebraska, 19, 80, 91, 94, 106, 123
Todd, Shellia, 14
Tonight Show, 10
Townsend, Elizabeth, 148
Tracy, H.H., 92, 115, 139
Twain, Mark, 13, 68

Union Creek, 97, 108, 117, 120
Union Pacific Railroad, 66
    damaged by floods, 50, 53–54, 86, 153
    in Johnny Carson's Johnny Comes Home, 10
    wrongful death lawsuit, 150
United Nations Intergovernmental Panel on Climate Change, 160
University of Nebraska, 5, 88, 104
unnatural disaster, xvi, 14–15, 43, 58, 102, 150, 152, 157–58, 160
Upper Elkhorn Natural Resource District, 130
US Army Corps of Engineers, xiii–xiv, 89–90, 102, 123, 141
    comprehensive improvement plans for Elkhorn River, 108–9, 112–14, 136
    flood control, 67, 75, 82, 105
    flood damage estimates for Nebraska (1940–1965), 128
    flood protection for Hannibal, Missouri, 13–15
    local protection projects for Elkhorn River Basin, 109–12, 116–17, 125, 146–47
    Missouri River office, 105
    navigation on the Mississippi River, 68
    response to Elkhorn flood of 1944, 104–16, 159
US Bureau of Reclamation, 68, 105
US Department of Agriculture, 88–89
US Engineer Office, 94
US Geological Survey, 25
US government river management policy, xv, 13–15, 82, 88, 108–15, 140, 158–59
US House of Representatives Flood Control Committee, 105–6
US National Wild and Scenic Rivers System, 12
US Route 20, 120
US Route 77, 101, 103
US Route 275, 5, 87, 96–98, 109, 120, 124–25, 145

US Soil Conservation District, 135
US Weather Bureau, 104

Valley, Nebraska, 130, 132
Vandelay, Art, 13
Vermillion, Dakota Territory, 31
Vulcraft, 145

Washington County, Nebraska, 19, 133, 142
Waterloo, Nebraska, 19, 53–54, 97, 109, 111, 116–17, 123, 132, 141
Wayne County, Nebraska, 90
Webb, Walter Prescott, 11, 22
Wells Mill, 56, 62
West, Elliott, 139
West Point Community Club, 88, 108
West Point Electric Light Company, 73
West Point, Nebraska, 19, 39–40, 51, 109, 111, 116–17, 122–23, 133
- flood damage estimate for 1940–1965, 128
- flood of 1881, 53. *See also* Elkhorn River, flood of 1881
- flood of 1891, 55–57. *See also* Elkhorn River, floods of 1891
- flood of 1912, 73. *See also* Elkhorn River, flood of 1912
- flood of 1920, 79–81. *See also* Elkhorn River, flood of 1920
- flood of 1940, 86–87. *See also* Elkhorn River, flood of 1940
- flood of 1944, 96–98. *See also* Elkhorn River, flood of 1944
- flood of 2010, 145–46. *See also* Elkhorn River, flood of 2010
- Grace Lutheran Church, 134
- millworks, 45–46, 50, 52. *See also* Elkhorn River, milling

West Point Manufacturing Company, 45
West Point Republican, 73, 79, 81
Western Sarpy County Drainage District, 68
Whetstone River, 30
Wildcat Creek, xiii–xiv
Willcocks, Sir William, 74–75
wind, 11–12, 20, 25–26, 29–30, 35, 42, 121,
Winslow, Nebraska, 19, 73, 87, 97, 103, 109, 111, 114–15, 123, 133
Wisner, Nebraska, 19, 72, 86, 97, 124, 146
WJAG radio station, 103
Wolfe, Emanuel, 7, 9, 61–62, 69–70
Woodland tradition, 37
Works Progress Administration, 82, 88–89
Worster, Donald, xv, 115–16, 154–57
Wunder, John, 5

Yellow River, 27
Young, Russell, 106

Zellmer, Sandra B., 14–15, 159–60
Zoroastrians, 27

ABOUT THE AUTHOR

Todd M. Kerstetter is an associate professor in the department of history at TCU. He is the author of *Inspiration and Innovation: Religion in the American West* and *God's Country, Uncle Sam's Land: Faith and Conflict in the American West.*